Informatik-Fachberichte 262

Herausgeber: W. Brauer
im Auftrag der Gesellschaft für Informatik (GI)

W. Gerth P. Baacke (Hrsg.)

PEARL 90 – Workshop über Realzeitsysteme

11. Fachtagung des PEARL-Vereins e.V.
unter Mitwirkung von GI und GMA
Boppard, 29./30. November 1990
Proceedings

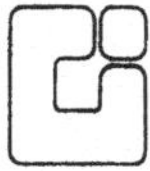

Springer-Verlag

Berlin Heidelberg New York London
Paris Tokyo Hong Kong Barcelona

Herausgeber

Wilfried Gerth
Institut für Regelungstechnik, Universität Hannover
Appelstr. 11, W-3000 Hannover

Per Baacke
Institut für Automatisierungstechnik, Universität Bremen, NW 1/N 1171
Postfach 33 04 40, W-2800 Bremen 33

CR Subject Classification (1987): C.3, C.2.5, D.2.1-2, D.3.2

ISBN-13: 978-3-540-53464-8   e-ISBN-13: 978-3-642-46725-7
DOI: 10.1007/978-3-642-46725-7

2145/3140-543210 – Gedruckt auf säurefreiem Papier

Vorwort

Die 11. Fachtagung des PEARL-Vereins zum Themenumfeld der
Realzeitprogrammierung findet unter neuen Randbedingungen
statt. Mitte dieses Jahres wurde nämlich erstmalig der
Kongreß 'Echtzeit 90' mit begleitender Ausstellung ver-
anstaltet, eine dreitägige große Tagung mit Parallelsitzun-
gen. In der Liste des Kongreßbeirates standen etliche Namen,
die in den letzten Jahren bei der Ausrichtung dieses
PEARL-Workshops mit dabei waren. Das ist natürlich kein
Zufall - schließlich unterstützte der PEARL-Verein tatkräf-
tig das Kongreß-Experiment und wirkte bei der Realisierung
mit.

In dieser Lage mußte man sich selbstverständlich Gedanken
machen, ob diese sehr groß geratene Veranstaltung im Juni
überhaupt noch Platz lassen würde für die heutige Fachtagung
sowie den eigentlich für September geplanten 'Personal-
Realtime-Computing'-Workshop. Als Konsequenz aus der neuen
Situation wurde der letztgenannte Workshop aufgegeben, und
fortan wird diese Fachtagung auch seine wichtigsten Funktionen mit übernehmen. Nun dürfen wir hoffen, daß viele Kollegen
aus Wissenschaft und Industrie auch weiterhin im kleineren
Kreis zu dieser Fachtagung zusammenkommen möchten.

Inhaltlich ist diese Fachtagung zwar stärker als andere dem
PEARL-Umfeld zugeordnet, doch wie jedes Jahr, so war auch
diesmal ein Blick über den Zaun durch hierfür geeignete
Beiträge und Diskussionen eine zwingende Vorgabe. Dies gilt
besonders für allgemeine Trends und neue Standards - sowohl
bei der Projektierung als auch bei der Realisierung von
Echtzeitsystemen. Der Kongreß 'Echtzeit 90' lieferte den
Teilnehmern ein Bild von der Konkurrenz verschiedener Hard-
und Softwarewerkzeuge. Darüber hinaus wird wohl auch eine
Expansion der Methoden zur objektorientierten Programmierung
in den Bereich der Echtzeitsysteme hinein zukünftig weitere
Vielfalt bringen.

Wenn wir unser Augenmerk hier einmal auf heutige Möglichkeiten der programmtechnischen Realisierung von Echtzeitsystemen richten, so bot der Juni-Kongreß dafür vier wesentliche Hauptlinien an:

- die Realisierung in C und Verwendung einer Echtzeitbiblio-
  theksschnittstelle (RTEID, ORKID, etc.) zu irgendeinem
  Echtzeitbetriebssystem. In diesem Umfeld gibt es auch
  verschiedene eigenartige Anläufe in Richtung auf Realtime-
  Unix-Derivate.

- die Realisierung in Ada und Betrachtung der Inter-
  ruptprozesse für Zeitmarken und externe Ereignisse als
  'Hardwaretasks', die als Partner für Rendezvous dienen. Ob
  ein und ggf. welches Laufbetriebssystem benötigt wird,
  ist anscheinend sehr uneinheitlich.

- bei übersichtlichen Kleinprojekten verwenden FORTH-Anhänger
  Echtzeit-Derivate von FORTH.

- Einsatz der Sprache PEARL, in der alle Echtzeitanweisungen
  auf Hochsprachebene compilerprüfbar und unabhängig vom
  benutzten unterlagerten Betriebssystem portabel definiert
  sind. Um die arteigene Schnittstelle des aktuellen
  Betriebssystems kümmert der Anwender sich überhaupt nicht.

Die Verwendung anderer Sprachen wie Assembler, FORTRAN,
Pascal, Modula etc. mit irgendwelchen nicht standardisierten
Echtzeitprozeduren stagniert anscheinend. Nach meinem Kon-
greßeindruck hat PEARL als bewährte und in sich schlüssige
Lösung eine gute Position. Doch obwohl Konzept, Erfahrungen
und Verfügbarkeit gut sind, ist die Wettbewerbslage aus
anderen Gründen schwierig. Um überhaupt bestehen zu können,
muß der Ansatz ständig an Hand neuerer Erkenntnisse und mit
Blick auf die konkurrierenden Lösungen weiterentwickelt werden - freilich ohne Preisgabe der Aufwärtskompatibilität.

Für die heutige Tagung wurde eine erfreulich hohe Zahl von
Beiträgen eingereicht. Zwangsläufig konnte beinahe die Hälfte
von ihnen am Ende nicht berücksichtigt werden. Manche der
Zurückweisungen waren schmerzlich, weil es sich um interes-
sante Beiträge handelte, die aber entweder keinem der fünf
Schwerpunkte zuzuordnen waren oder aber zu einem bereits
überbuchten Bereich gehörten. Die in diesem Band gesammelten
Beiträge haben wir folgenden Themengruppen zugeordnet:

- Konzepte und theoretische Aspekte der Realzeit-DV.
- Software Engineering in Realzeitprojekten.
- Fertigungsautomatisierung.
- Realzeit-Anwendungen.
- Echtzeit-Standards.

Wir hoffen, insgesamt einen interessanten wissenschaftlichen
Mix aus Theorie, Methodik und Anwendungserfahrungen erzielt
zu haben. Dabei konnte leider nur noch Platz für drei
Beiträge zur Beschreibung praktischer PEARL-Anwendungen vorgesehen werden. Ganz gezielt wurde darum genau je ein Beitrag
für die drei Stufen 'kleinst', 'mittel' und 'groß' beim
Kapitalvolumen im Anwendungsumfeld eingeplant.

Wir danken allen Vortragenden für Ihre Mühen und die Bereitschaft, den Teilnehmern wertvolle Hinweise und Erfahrungen zu vermitteln. Dabei schließen wir ausdrücklich auch jene
ein, die diesmal nicht berücksichtigt werden konnten. Ein
besonderer Dank gilt auch dem Springer-Verlag, der durch

seine Bereitschaft zur Veröffentlichung des Tagungsbandes sicher ganz wesentlich zur Attraktivität und Öffentlichkeitswirkung der Tagung beigetragen hat; bereits der Tagungsband zu PEARL 89 erschien als Band 231 der Informatik-Fachberichte. Herrn Dipl.-Ing. Baacke vom PEARL-Verein gebührt Dank und Anerkennung für die organisatorische Abwicklung.

Denken wir stets daran, daß jemand, der in einer uns nicht bekannten oder von uns verworfenen Sprach- bzw. Methodenwelt Realzeitsysteme projektiert und realisiert, seine Methode nicht deswegen favorisiert, weil er uns damit ärgern will. Bei allem Engagement ist Aggressivität also wirklich fehl am Platze. Hüten wir uns auch vor der Arroganz, im Besitz der einzig wahren Echtzeitlehre zu sein - solche geistige Bequemlichkeit überzeugt niemanden. Eine möglichst objektive gegenseitige Information kann dagegen viele Brücken schlagen und beide Seiten bereichern, selbst wenn jeder auf seinem eigenen Wege bleibt - neu bestätigt oder nachdenklich - in jedem Falle aber klüger als vorher. Möge diese Fachtagung auch in diesem Sinne wiederum erfolgreich sein! Stellvertretend für das Programmkomitee und für den Veranstalter wünsche ich allen Teilnehmern interessante Anregungen, Diskussionen und neue Erkenntnisse.

Hannover, im September 1990   W. Gerth

P r o g r a m m k o m i t e e

D. Eberitzsch	Bremen
W. Gerth	Hannover (Vorsitz)
W. Halang	Groningen
R. Henn	Oberhaching
A. Küchle	Friedrichshafen
K. Mangold	Konstanz
D. Popovic	Bremen
H. Rzehak	Neubiberg
K. Stieger	Neubiberg
H. Windauer	Lüneburg

Inhaltsverzeichnis

Realzeit-Anwendungen

Echtzeit-Standards

Real-Time Euclid: Concepts Useful for the Further Development of PEARL

Alexander D. Stoyenko

Department of Computer and Information Science

New Jersey Institute of Technology

Newark, NJ 07102

U.S.A.

alex@vienna.njit.edu

Abstract

PEARL is a widely-used, standardised, industrial process-control programming language. The programs, that are written in PEARL, are typically expected to operate in a predictable, timely, reliable manner. Developing such programs is no easy task. Yet, with the invent of the experimental programming language Real-Time Euclid and its *schedulability analyser*, a tool that is used to predict whether real-time software will adhere to its critical timing constraints, it has been demonstrated that this task may be made considerably easier.

In this paper, we describe how PEARL may be modified to be better suitable for real-time application writing, and to accommodate schedulability analysis. We aim to contribute to the current revision of PEARL, and define a new standard we refer to as *High-Integrity (HI) PEARL*. Building on the powerful foundation of the existing PEARL definition, and further enhanced with the mechanisms introduced by Real-Time Euclid, HI-PEARL should be the top choice of the real-time programmer of the 90's and the decades to follow.

1   Introduction

The requirements of real-time applications define a paradigm of computing that is very different from traditional computing paradigms. While in a traditional paradigm, such as interactive processing, for instance, the correctness of a program is independent of the timing characteristics of its execution, a real-time program must by its very nature adhere to the timing constraints of its application. More completely, real-time operation of a computer system may be defined as "The operating mode of a computer system in which the programs for the processing of data arriving from the outside are permanently ready, so that their results will be available within predetermined periods of time; the arrival times of the data can be randomly distributed or be already a priori determined depending upon the different applications." [1].

Hundreds of languages have been designed to address not only such main traditional paradigms of computing as batch or interactive, but even specific areas within the paradigms. Thus, FORTRAN [2] has been designed specifically for scientific computing, COBOL [3] for business applications and so forth. Yet, in the past, when a language has been designed or designated to be used in real-time computing, it has lacked, to a significant extent or entirely, the notion of real-time as a first-class entity. Consider, for instance, Ada [4]: a language designed to address the domain of embedded real-time applications. The language is well-structured, strongly typed, has adequate exception handlers, an elegant and clever synchronisation mechanism (the rendezvous), supports multi-tasking, and is, in general, well-designed. Yet, Ada has no constructs to allow the programmer to express such basic timing constraints as deadlines or process durations.

The lack of adequate real-time languages cannot be explained away by a lack of applications. On the contrary, real-time applications are widespread and their number is fast growing. To name just very few, these include windtunnel control [5], astronomical data acquisition [6], plane control [7], space shuttle control [8], overcurrent and high impedance fault relaying [9], radar applications [10], and amusement park ride control [11]. Rather, conventional or nearly conventional languages have been used simply because of their wide availability and user's familiarity with them.

Naturally, real-time software can be written in any available language. By the same argument, one can write scientific software in COBOL, implementing high-precision floating-point arithmetic from first principles. If all languages available to the scientific programmer did in fact lack floating-point arithmetic, the programmers would then have to implement it from first principles. Mercifully, FORTRAN and other scientific languages are available. Yet, a real-time language is not. Thus, real-time programmers are implementing timing constraints and operations from first principles.

So far, we have discussed the availability of adequate real-time languages for the industrial programmer. The situation in the real-time language scene is inadequate and no such language is available [12]. However, a number of real-time languages have been designed at various universities. The first language design that has forced every program to express its timing constraints, defined a real-time process model, and forbade or restricted all constructs that take arbitrarily-long to execute was Real-Time Euclid [13, 14]. Real-Time Euclid has also included a new compile-time tool, called by the author *a schedulability analyser*, which predicts whether a program will meet its timing constraints [15, 14, 16]. Real-Time Euclid was the first language which guaranteed that its every program was schedulability analysable. A single prototype of the language and its schedulability analyser has been implemented and validated. A number of other efforts followed Real-Time Euclid, such as [17–20] to name just a few.

2 Requirements of a Real-Time Language

To facilitate real-time program writing, a language has to satisfy a number of requirements.

2.1 Timing Constraint Expressibility and Schedulability Analysability

Timing constraints must be explicitly expressible in the language. A real-time program should be designed and implemented as a set of real-time tasks each controlling or observing a corresponding real-time process of the application.

At least a usable subset of the language should be schedulability analysable. This means, that every program written in the subset can be statically analysed to ascertain whether it will meet its timing constraints. The following discussion is intended to motivate the requirement of schedulability analysability.

A real-time application imposes critical timing constraints on the software that controls it. The time-critical nature of a real-time application makes it necessary that to be considered reliable and predictable, the controlling software must adhere predictably to these constraints. The timing constraints are typically described in terms of timing schedules that govern task activation, execution and termination. Each task activation is expected to perform its computations successfully before a deadline, explicitly specified in its schedule, expires.

Since deadlines and other such timing constraints directly correspond to explicit application requirements, the computation model of a task activation has no need to include any abstractions that may prevent the execution from adhering to the schedule. Thus, any abstraction that takes arbitrarily long to execute is unnecessary. From the theory of decidability of computation models, it follows that for any analysis, that attempts to ascertain a priori the adherence of a set of real-time task executions to their schedules, to succeed, any computation abstraction that takes arbitrarily long to execute must, in fact, be forbidden.

If we apply this reasoning to a high-level real-time language, then it follows that the language should allow the expression of timing schedules and real-time tasks whose activation, execution and termination are controlled by schedules. On the other hand, the language should not allow any construct that may take arbitrarily long to execute to be used in task construction.

Thus, a number of the familiar conventional constructs, such as, for instance, a general while-loop, should be banned from real-time languages. Indeed, an iterative computation that has a deadline must be expressible in terms of a constant-bounded loop. Otherwise, the computation may miss its deadline and thus fail. If, on the other hand, unbounded iteration is called for in the real-time application, then it must be the case that the iterative computation has no inherent deadline associated with it. Then, a proper way to implement the computation is to make every single iteration (or a bounded number of iterations) a task activation.

Once a language is modified to describe each program as a set of real-time tasks whose activation, execution and termination is governed by explicit schedules, and to forbid any constructs that may take arbitrarily long to execute, expressing real-time applications in the language becomes quite natural. Moreover, every program in the language becomes schedulability analysable.

2.2 Other Requirements

Real-time software must be very reliable in general. Thus, a real-time language should be very secure, in the sense of reliability and robustness. The language must make provisions for exception handling. The language should also have strong typing, structured constructs and be modular.

Real-time software almost always involves multiprogramming. Thus, a real-time language should include task definitions, as well as structured task synchronisation mechanisms.

Many real-time systems are quite large. For instance, [21] reports a survey of twenty projects undertaken at eleven companies. On the average, a project involved 30 programmers coding 150,000 instructions. Under such circumstances, we believe that a real-time language should be very modular and allow separate compilation, thus scaling up well for "programming-in-the-large".

Real-time software typically interfaces with many different conventional and special purpose hardware devices. It is important, therefore, to provide constructs in the language to access hardware locations and handle interrupts directly in a flexible and yet secure way.

Real-time systems have a long life expectancy, because so do the environments they control. It is well-known that the majority of the costs associated with such systems are maintenance costs. We therefore require that real-time programs should be easy to understand, read and modify. This general maintainability requirement is greatly aided if the language is readable, well-structured and not too complex.

3   PEARL

Of the languages available to the industrial real-time programmer, *P*rocess and *E*xperiment *A*uto-
mation *R*ealtime *L*anguage, or *PEARL* [22] comes closest to meeting the requirements of a real-time
language. Although conceived a decade earlier then Ada, PEARL provides more expressive power
for formulating real-time and process control applications than does Ada or any other language.
The reason may be that the language has been defined by electrical, chemical, and control systems
engineers, on the basis of their practical experience in industrial automation problems and devoting
special attention to the timing aspects of the applications. Sponsored by the Ministry of Research
and Technology of the Federal Republic of Germany, the development of PEARL has begun in the
late Sixties. There are two versions of PEARL: Basic PEARL and the superset Full PEARL. Both
versions have been standardised nationally, through DIN, and internationally, through ISO.

3.1   Single-System PEARL

PEARL includes all conventional elements of Pascal and Pascal-like languages, and extends the
basic data types with the additional data types CLOCK and DURATION, and their corresponding
operations.

To support modular construction of complex software systems, PEARL programs are composed of
separately-compilable modules. A module may contain a system division and a number of problem
divisions. A system division encapsulates and isolates hardware and other environment dependen-
cies. Such a division contains complete information necessary to run the module, relinquishing
the need to describe the environment in a job control language or otherwise outside of PEARL.
User-defined identifiers are associated with hardware devices and addresses in the system division.
These identifiers and not the actual devices are referenced in the consequent problem divisions.
Problem divisions are thus dedicated to expressing the algorithmic logic of the program, in a way
that is environment-independent.

Besides high-level file-oriented I/O statements, PEARL also provides low-level constructs to exchange information with process peripherals. These constructs do not communicate with the
peripherals directly, but rather go through virtual data station or DATIONs, associated with the
actual peripherals in the system divisions.

PEARL features a comprehensive, application-oriented range of features supporting task scheduling
and control, and the expression of time-constrained behaviour. In particular, there are simple
schedules depending on temporal events or interrupt occurrences, as well as general, cyclically
repetitive ones. Five statements are provided for the control of task state transitions. Their actual
execution can be linked to schedules. A task is transferred from the dormant to the ready state by
an activation, in the course of which the task's priority may be changed. The operation inverse to
this is the termination. Running or ready tasks can temporarily be suspended and later resumed.
Finally, all scheduled future activations of a task may be annihilated.

PEARL combines the advantages of general purpose and of special purpose synchronous real-time
languages [23]. The language avoids such Ada's problems as non-determinism in synchronisation
and communication, and lack of means for immediate task abortion.

3.2   Distributed-System PEARL

Recently a third part of PEARL has been standardised in Germany [24], which allows for the programming of distributed applications and for dynamic system reconfiguration. The Multiprocessor-

PEARL language extensions include elements for describing the hardware configuration of a multiprocessor system. Among these, the nodes or STATIONs and the external peripherals are described, as well as the physical communication network that connects them. The distribution of software units among the nodes is described, as are the logical communication channels, transmission protocols, and failure recovery procedures.

Multiprocessor-PEARL programs are structured by grouping modules in COLLECTIONs. The collections are distributed statically or dynamically among system nodes. Since collections may move among nodes, system divisions may not be contained in a collection. Instead, a single system division is provided for each station in a given configuration. In addition, there is a global system division to enable the access to devices attached to other nodes. To allow dynamic reconfiguration in response to system state changes, Multiprocessor-PEARL provides configuration divisions and corresponding executable statements for loading and unloading of collections and for the establishment and the discontinuation of logical communication paths.

Communication among collections is performed through message exchange. The message exchange avoids referencing the communication objects in other collections directly, and decouples the communication structure from the logic of message passing. The exchange is supported through communication endpoints or PORTs. Ports represent interfaces between collections and the outside world. A collection may have an arbitrary number of them. There are input and output ports. In problem divisions of collections, messages are routed through the ports. The logical communication paths among collections in a certain configuration are established by executing language statements to connect ports. One-to-many and many-to-one communication structures may be set up. Logical transmission links may be mapped to physical links in three different ways: by selecting specific physical links, by specifying preferred physical links, or by leaving the selection to the network operating system. A message may be sent using one of three protocols: the asynchronous "no-wait-send", the synchronous "blocking-send", or the synchronous "send-reply". If a synchronous send or receive operation is selected, it may be executed with a timeout clause.

3.3 PEARL's Limitations

Nevertheless, PEARL has a number of shortcomings. One significant limitation is the lack of well-structured synchronisation primitives with temporal supervision. PEARL's exception handling facility is unstructured. The language lacks the means to interrogate task and resource states. It is possible to construct a PEARL program that will take arbitrarily long to execute, in a single task activation. Thus, the language makes insufficient provisions to enable schedulability analysis of its programs.

4 Real-Time Euclid

A descendant of the Pascal [25], Euclid [26], and Concurrent Euclid [27] line of languages, Real-Time Euclid has been presented in detail elsewhere [13, 14]. A short description suffices for the purposes of this presentation.

Real-Time Euclid has been designed with a sufficient set of provisions for schedulability analysis. Thus, every Real-Time Euclid program can be analysed at compile time to determine whether or not it will guarantee to meet all time deadlines during execution. The language has processes. Each process is associated with a frame (a minimal period), and can be activated periodically, by a signal, or at a specific compile-time specified time. Once activated, a process must complete its task before the end of the current frame. After activation, a process cannot be reactivated until the end of the current frame.

Real-Time Euclid has no constructs that can take arbitrarily long to execute. The only loops allowed are constant-count loops. Recursion and dynamic variables are disallowed. Wait- and device-condition variables time out if no signal is received during a specified time delay. Process synchronisation is achieved through monitors [28], waits, signals, outside-monitor waits and broadcasts. Dynamic monitor nesting is supported.

The language has structured, time-bounded exception handlers. The handlers are allowed in any procedure, function or process. Exception propagation is single-thread, in the opposite direction to a chain of calls. There are three classes of exceptions, based on their severity.

Real-Time Euclid is modular, procedural, and strongly-typed. It is structured and small enough to be remembered in its entirety, thus facilitating programming-in-the-small. Modularity and separate compilation make Real-Time Euclid a suitable language for programming-in-the-large.

5 Modifying PEARL

A detailed description of how PEARL may be modified to benefit from the concepts introduced by Real-Time Euclid and by a number of other earlier suggested modifications [29–31] are forthcoming [32, 33]. We now present an outline of the PEARL modifications.

To enable upward compatibility of PEARL programs, and to minimise the impact on the PEARL user community, all of the following are optional. In essence, these define a schedulability analysable dialect of PEARL, or a schedulability analysable subset of the new PEARL.

All temporal conditions used in a schedule must be in terms of compile-time constants. Every task is associated with a frame size, expressed as a compile-time constant. Once a task is activated, it must complete by the time computed as the sum of the current time and the frame size. Usually, the scheduler will detect well in advance whether the task will be able to complete by that time, that is, its deadline. Special signals are raised if this is impossible or the deadline is actually missed. These restrictions ensure that the schedules are only used to express explicit timing constraints, as imposed by the applications.

Semaphores are removed from the language and replaced by the structured **LOCK** mechanism for the formulation of critical regions. Endowed with a timeout constant to bound the queue waiting time, an alternative action to be taken in the case of unsuccessful waiting, and an execution time bound for the body of the **LOCK**, it is possible to determine, already at compile-time, a **NOLONGERTHAN** parameter for each **LOCK** statement. The I/O operations, namely **PUT**, **GET**, **READ**, **WRITE**, **SEND** and **TAKE**, are realised as runtime system procedures with known upper execution time bounds. When accessing devices shared among several tasks, these must be used within the scopes of **LOCK** statements. The above provisions ensure that no synchronisation or I/O operation takes arbitrarily long to execute.

Every signal event declaration is associated with a compile-time limit on the number of times the event may be raised in a single activation of a task causing and handling the exception. The same limit applies, naturally, to all specifications (imports) of the same event throughout the program. The minimum duration between two subsequent occurrences of the same interrupt, which is equivalent to its maximum arrival frequency, is specified in the interrupt declaration. These restrictions ensure that there can be no arbitrarily long chains of raised signals or interrupts.

To prevent unpredictable iteration, **WHILE** clauses are no longer permitted in **REPEAT** statements. Moreover, every **REPEAT** must specify a maximum compile-time number of iterations. The **GOTO** statement is removed from the language. To avoid arbitrarily long chains of invocations, (mutually) recursive calls of PEARL procedures are no longer allowed.

Blocking communication operations are associated with **NOLONGERTHAN** clauses. This ensures that no communication delay may be arbitrary.

To enhance the use of existing, non-schedulability-analysable PEARL constructs, compiler pragmas are added to specify and enforce timing constraints. For instance, **REPEAT** with a **WHILE** clause may be associated with a pragma that will say {**NOMORETHAN n ITERATIONS**} and so forth. The pragmas enable the schedulability analyser to produce partial schedulability information for a non-analysable program. Thus, in our example, while the analyser may be unable to predict guaranteed response times for the entire program, it may be able to put an upper time bound on the **REPEAT**.

A schedulability analyser is provided for PEARL [33]. The analyser consists of a front end and a back end. The front end is incorporated into the compiler, and is charged with extracting compile-unit timing information. The back end is a standalone machine-independent tool, that combines the information extracted by the front end, assesses such intermediate metrics as task contention and other delays, and derives worst-case schedulability metrics, such as guaranteed response times.

Many schedulability analysis techniques reported in the literature [34–45] have dealt with resource queuing by means of solutions that do not take into account segments' relative positions and inter-segment distances on the time line [35] when estimating task contention and other delays. These solutions are typically closed-form or polynomial, and thus work very quickly. However, they result in overly pessimistic worst-case delays, and ultimately in overly pessimistic guaranteed response times [46].

In the back end of the analyser, we use *frame superimposition* [14] to derive intermediate and ultimate schedulability metrics. Frame superimposition simply means fixing a single task's frame at its starting time and positioning frames of other tasks along the time line in such a way as to maximise the amount of resource contention the task incurs. Frame superimposition uses relative positions of segments and distances between segments on the time line. The algorithm shifts frames exhaustively, for every time unit, for every task, for every combination of frames possible.

The frame superimposition algorithm is clearly exponential. In fact, finding the optimal worst-case bound for resource contention in the presence of deadlines is NP-complete, given that even the most basic deadline scheduling problems are [47, 48]. However, this part of the analysis operates on a small number of objects (segments, each combining a great many statements), in a relatively small number of arrangements. Moreover, it has been demonstrated in the Real-Time Euclid evaluation that considerably better delay bounds are derived by our algorithm, than by closed-form or polynomial algorithms [14, 49, 50]. Thus, we feel justified in using frame superimposition. Nevertheless, we are working on improving the efficiency of frame superimposition [51].

6 Concluding Remarks

We have argued for making PEARL a language even more suitable for real-time programming than it already is. Obviously, we feel that introducing such Real-Time Euclid concepts as explicit timing constraint expressions (already present to a noticeable extent in PEARL), defining an optional schedulability analysable subset of the new PEARL and introducing optional schedulability pragmas for the remaining subset is of paramount importance to the further development and acceptance of the language. As FORTRAN without floating-point operations is inconceivable as a scientific language, so too any language without the notions of real-time and schedulability analysis built-in is not a real-time language.

The question then is, what would happen if PEARL were to develop without including these real-time features? Would the German and international industrial process applications collapse? Would the programmers charged with real-time software writing go on strike? Clearly, the answer to these and other such questions is "most certainly not!".

Why is it then that we insist on introducing these new features and defining HI-PEARL? We do this not to define some abstract, binary must: either this or a failure. Quite the contrary, there are few absolute musts, if any, in the software or computer business. We insist on the new definition for PEARL to *aid significantly* in the development process of high-integrity, real-time software. We refuse to continue making the programmer implement real-time mechanisms from first principles. Enough of substituting loop iterations for task frames. Enough of hand-measuring upper execution times of code modules. It is time to face reality and introduce real-time into real-time languages.

The only remaining question then is, why have we picked PEARL and not some other language? The answer to this is quite simple. PEARL is both the only standardised industrial-control language in wide use, and the one that already comes closest to meeting the requirements of a real-time language.

PEARL, hopefully through HI-PEARL, has a bright future ahead of it. Already there is work on standardising the world's first provably correct real-time computer [52], on which the world's first provably correct real-time kernel software [53] will be built. HI-PEARL will complement this and any other platform as the world's first standardised, widely-used, truly real-time language.

7   Acknowledgements

I thank a great number of people who contributed to the ideas that made it into this paper. Many thanks go to Gene Kligerman, Chris Ngan, Gerry Parnis, Scott Thurlow, Greg Nymich and Victor Anderson, the members of the original Real-Time Euclid Project. Very special thanks and gratitude go to Wolfgang Halang whose extraordinary vision and contribution to the field of real-time computing are an inspiration to all of us, and whose help made much of my recent work possible. I am deeply in debt to Prof. Popovic and other PEARL'90 programme committee members for inviting me to address this distinguished assembly.

References

[1] *DIN 44 300, Informationsverarbeitung*, Beuth-Verlag, Berlin, 1985.

[2] IBM Corporation, *Specifications for the IBM Mathematical FORmula TRANslating System, FORTRAN*, New York, November 1954.

[3] U.S. Department of Defence, *COBOL, Initial Specifications for a Common Business Oriented Language*, 1960.

[4] *The Programming Language Ada Reference Manual*, American National Standards Institute, Inc., ANSI/MIL-STD-1815A-1983. Lecture Notes in Computer Science 155, Springer-Verlag, Berlin-Heidelberg-New York-Tokyo, 1983.

[5] C. D. Williams, "The Data Acquisition, Data Reduction and Control System (DARCS) for the NRCC 2x3m Windtunnel," *Proceedings of the IEEE 1984 Real-Time Systems Symposium*, December 1984, pp. 89 – 94.

[6] P. W. Kelton, "Distributed Computing for Astronomical Data Acquisition at McDonald Observatory," *Proceedings of the IEEE 1984 Real-Time Systems Symposium*, December 1984, pp. 83 – 88.

[7] G. Kaplan, "The X-29: Is it coming or going?," *IEEE Spectrum*, June 1985, pp. 54 – 60.

[8] G. D. Carlow, "Architecture of the Space Shuttle Primary Avionics Software System," *Communications of the ACM*, Vol. 27, No. 9, September 1984, pp. 926 – 936.

[9] B. M. Aucoin, R. P. Heller, "Overcurrent and High Impedance Fault Relaying using a Microcomputer," *Proceedings of the 7th Texas Conference on Computing Systems*, November 1978, pp. 2.5 – 2.9.

[10] E. T. Fathi, N. R. Fines, "Real-Time Data Acquisition, Processing and Distribution for Radar Applications," *Proceedings of the IEEE 1984 Real-Time Systems Symposium*, December 1984, pp. 95 – 101.

[11] V. P. Nelson, H. L. Fellows, Jr., "A Microcomputer-Based Controller for an Amusement Park Ride," *IEEE Micro*, August 1981, pp. 13 – 22.

[12] W. A. Halang, A. D. Stoyenko, "Comparative Evaluation of High-Level Real-Time Programming Languages," to appear in *International Journal of Real-Time Systems*, 1990.

[13] A. D. Stoyenko, E. Kligerman, "Real-Time Euclid: A Language for Reliable Real-Time Systems," *IEEE Transactions on Software Engineering*, Volume 12, Number 9, September 1986. Also in *Hard-Real-Time Systems*, edited by J. Stankovic and K. Ramamritham, IEEE Press, Maryland, 1988.

[14] A. D. Stoyenko, *A Real-Time Language With A Schedulability Analyzer*, Ph.D. Thesis, University of Toronto, 1987. Also available as Computer Systems Research Institute, University of Toronto, Technical Report CSRI-206.

[15] A. D. Stoyenko, "A Schedulability Analyzer for Real-Time Euclid," *IEEE 1987 Real-Time Systems Symposium*, San Jose, California.

[16] A. D. Stoyenko, V. C. Hamacher, R. C. Holt, "Schedulability Analysis of Hard-Real-Time Programs," second revision, submitted in 1987 to *IEEE Transactions on Software Engineering*.

[17] J. W. S. Liu, S. Natarjan, "Expressing and Maintaining Timing Constraints in FLEX," *Proceedings of the IEEE 1988 Real-Time Systems Symposium*, December 1988.

[18] M. Donner, D. Jameson, "Language and Operating Systems Features for Real-Time Programming," *Computing Systems*, Winter 1988.

[19] Y. Ishikawa, H. Tokuda, C. W. Mercer, *Object-Oriented Real-Time Language Design: Constructs for Timing Constraints*, Department of Computer Science, Carnegie Mellon University, Technical Report CMU-CS-90-111, March 1990.

[20] A. Shaw, *Reasoning About Time in Higher-Level Language Software*, Department of Computer Science, University of Washington, Technical Report 87 – 08 – 05, August 1987.

[21] R. L. Glass, "Real-Time: The "Lost World" of Software Debugging and Testing," *Communications of the ACM*, Vol. 23, No. 5, pp. 264 – 271, May 1980.

[22] DIN 66 253, *Programming Language PEARL (Programmiersprache PEARL)*, Teil 1 Basic PEARL, 1981; Teil 2 Full PEARL, Beuth-Verlag, Berlin, 1982.

[23] G. Berry, "Real Time Programming: Special Purpose or General Purpose Languages," *Proceedings of the 11th IFIP World Computer Congress*, San Francisco, 1989.

[24] DIN 66 253, *Programmiersprache PEARL - Mehrrechner-PEARL*, Teil 3, Beuth-Verlag, Berlin, 1989.

[25] N. Wirth, "The Programming Language Pascal," *Acta Informatica 1*, 1971, pp. 35 – 63.

[26] R. C. Holt et al., "Euclid: A language for producing quality software," *Proceedings of the National Computer Conference*, Chicago, May 1981.

[27] J. R. Cordy, R. C. Holt, *Specification of Concurrent Euclid*, Technical Report CSRG-133, Computer Systems Research Group, University of Toronto, August 1981.

[28] C. A. R. Hoare, "Monitors: An Operating System Structuring Concept," *Communications of the ACM*, Vol. 17, No. 10, October 1974, pp. 549 – 557.

[29] W. A. Halang, "A Proposal for Extensions of PEARL to Facilitate the Formulation of Hard Real-Time Applications," *Informatik-Fachberichte 86*, pp. 573 – 582, Springer-Verlag, Berlin-Heidelberg-New York-Tokyo, 1984.

[30] W. A. Halang, R. Henn, "Additional PEARL Language Structures for the Implementation of Reliable and Inherently Safe Real-Time Systems," *Proceedings of the 15th IFAC/IFIP Workshop on Real Time Programming*, pp. 35 – 42, Pergamon Press, Oxford, 1988.

[31] W. Ehrenberger, "Softwarezuverlässigkeit und Programmiersprache," *PEARL-Rundschau*, Vol. 3, No. 2, 1982, pp. 49 – 55.

[32] A. D. Stoyenko, W. A. Halang, "A Proposal for High-Integrity (HI) PEARL: An Industrial Standard Language for High-Integrity Applications," *working paper*, 1990.

[33] A. D. Stoyenko, W. A. Halang, "Analysing PEARL Programs for Timely Executability and Schedulability" ("Analyse zeitgerechter Zuteilbarkeit und Ausführbarkeit von PEARL-Programmen"), to appear in *Proceedings of Fachtagung Prozessrechensysteme '91*, Berlin, Germany, February 1991.

[34] D. W. Leinbaugh, "Guaranteed Response Times in a Hard-Real-Time Environment," *IEEE Transactions on Software Engineering*, Vol. SE-6, No. 1, January 1980, pp. 85 – 91.

[35] D. W. Leinbaugh, M.-R. Yamini, "Guaranteed Response Times in a Distributed Hard-Real-Time Environment," *Proceedings of the IEEE 1982 Real-Time Systems Symposium*, December 1982, pp. 157 – 169.

[36] C. L. Liu, J. W. Layland, "Scheduling Algorithms for Multiprogramming in a Hard-Real-Time Environment," *JACM*, Vol. 20, No. 1, January 1973, pp. 46 – 61.

[37] A. K. Mok, "The Design of Real-Time Programming Systems Based on Process Models," *Proceedings of the IEEE 1984 Real-Time Systems Symposium*, December 1984, pp. 5 – 17.

[38] A. K. Mok, M. L. Dertouzos, "Multiprocessor Scheduling in a Hard-Real-Time Environment," *Proceedings of the 7th Texas Conference on Computing Systems*, November 1978, pp. 5.1 – 5.12.

[39] K. Ramamritham, J. A. Stankovic, "Dynamic Task Scheduling in Distributed Hard Real-Time Systems," *Proceedings of the IEEE 4th International Conference on Distributed Computing Systems*, May 1984, pp. 96 – 107.

[40] K. Ramamritham, J. A. Stankovic, S. Cheng, "Evaluation of a Flexible Task Scheduling Algorithm for Distributed Hard Real-Time Systems," *IEEE Transactions on Computers,* Vol. C-34, No. 12, December 1985, pp. 1130-1143.

[41] P. G. Sorenson, *A Methodology for Real-Time System Development,* Ph.D. Thesis, Department of Computer Science, University of Toronto, 1974.

[42] P. G. Sorenson, V. C. Hamacher, "A Real-Time System Design Methodology," *INFOR,* Vol. 13, No. 1, February 1975, pp. 1 – 18.

[43] A. D. Stoyenko, *Real-Time Systems: Scheduling and Structure,* M.Sc. Thesis, Department of Computer Science, University of Toronto, 1984.

[44] T. J. Teixeira, "Static Priority Interrupt Scheduling," *Proceedings of the 7th Texas Conference on Computing Systems,* November 1978, pp. 5.13 – 5.18.

[45] W. Zhao, K. Ramamritham, "Distributed Scheduling Using Bidding and Focused Addressing," *Proceedings of the IEEE 1985 Real-Time Systems Symposium,* December 1985, pp. 103 – 111.

[46] A. D. Stoyenko, *The State of the Art in Hard-Real-Time Modeling and Languages,* Ph.D. Qualifying Paper, Department of Computer Science, University of Toronto, May 1986.

[47] M. R. Garey, D. S. Johnson, "Complexity Results for Multiprocessor Scheduling under Resource Constraints," *SIAM Journal on Computing,* Vol. 4, No. 4, December 1975, pp. 397 – 411.

[48] J. D. Ullman, "Polynomial complete scheduling problems," *Proceedings of the 4th Symposium on OS Principles,* 1973, pp. 96 – 101.

[49] S. A. Thurlow, *Simulation of a Real Time Control System Using the Real-Time Euclid Programming Language,* Student Project Report, Department of Computer Science, University of Toronto, April 1987.

[50] G. Parnis, *Simulation of Packet Level Handshaking in X.25 Using the Real-Time Euclid Programming Language,* Student Project Report, Department of Computer Science, University of Toronto, April 1987.

[51] A. D. Stoyenko, T. Marlowe, "Analysis of Real-Time Programs Using Hierarchical Clustering," *working paper,* 1990.

[52] W. A. Halang, S.-K. Jung, "A Concept of a Computer System for the Execution of Safety Critical Licensable Software Programmed in a High Level Language," to appear in *Proceedings of the IFAC SAFECOMP '90 Conference,* London, U.K., October – November 1990.

[53] W. A. Halang, A. D. Stoyenko, *Constructing Predictable Real Time Systems,* Kluwer Academic Publishers, Boston-Dordrecht-London, 1991.

Erweiterung und Anwendung von PEARL zur Programmierung speicherprogrammierbarer Steuerungen

Wolfgang A. Halang

Reichsuniversität zu Groningen
Fachgruppe Informatik
Postfach 800
NL-9700 AV Groningen

Zusammenfassung

Um die führende Rolle von PEARL als *die* Sprache für den Automatisierungstechniker auch weiterhin zu gewährleisten, ist es unbedingt notwendig, auch Sprachmittel zur Formulierung sequentieller Ablaufsteuerungen in PEARL zur Verfügung zu stellen, da diese in den Anwendungen in letzter Zeit an Bedeutung gewinnen. Deshalb werden geeignete Sprachelemente, die sich problemlos in die bestehende PEARL-Syntax einfügen, definiert und in EBNF exakt beschrieben. Auf der Grundlage der Semantik sequentieller Abläufe wird gezeigt, dass sich diese Sprachmittel in Codesequenzen auflösen lassen, die in Standard-PEARL formuliert werden können. Mithin ist es nicht erforderlich, existierende PEARL-Compiler entsprechend zu erweitern, sondern es genügt die Bereitstellung eines Präprozessors zur Erkennung und Umformung der neuen Sprachelemente. Dadurch wird PEARL in die Lage versetzt, auch weiterhin das ganze Spektrum der in der Prozessautomatisierung benötigten Funktionalität bereitzustellen.

1. Einleitung

Unzweifelhaft steht PEARL, von der Leistungsfähigkeit und vom angebotenen Sprachumfang her gesehen, ohne ernstzunehmende Konkurrenz da. Dies gilt insbesondere im Hinblick auf die Eignung von PEARL für die Aufgaben der Automatisierungstechnik. Wie bereits im Einführungsvortrag zum letztjährigen PEARL-Workshop [1] ausgeführt, gibt es jedoch einen Typ von Echtzeitsystemen, der bisher von der Forschung und der Sprachentwicklung offensichtlich völlig unbeachtet geblieben ist: die *speicherprogrammierbaren Steuerungen*. Dieser Bereich der Prozessautomatisierung zeichnet sich durch recht heuristische Methoden aus. Die Programmierung der einzelnen Systeme erfolgt noch ausschliesslich in herstellerspezifischer Form und darüberhinaus auch auf recht niedrigem Niveau, nämlich in Assembler-ähnlichen textuellen Sprachen bzw. in Form graphischer Kontaktpläne. Um hier Abhilfe zu schaffen, erarbeitet die International Electrotechnical Commisson (IEC) zur Zeit eine internationale Norm [2], mit der eine Familie zweier graphischer und zweier textueller systemunabhängiger Programmiersprachen standardisiert werden soll. Diese vier Sprachen bauen auf gemeinsamen Elementen auf und sollen zueinander äquivalent und ineinander transformierbar sein. Jeweils eine graphische und eine textuelle Sprache haben schaltungsnahes bzw. maschinennahes Niveau. Die höhere graphische Sprache ist blockdiagrammorientiert und kommt der Denkweise des Ingenieurs zur Formulierung sequentieller Steuerungsabläufe und von Anwender-Software auf der Basis vorgefertigter Programmbausteine sehr entgegen.

Die im IEC-Entwurf beschriebene höhere textuelle Programmiersprache *Strukturierter Text (ST)* hat grosse Ähnlichkeit mit Pascal und erlaubt natürlich im Gegensatz zu den graphischen Sprachen eine uneingeschränkte Formulierbarkeit syntaktischer Details, der Algorithmik und der Kontrollstrukturen. Auf Grund ihres späteren Einsatzgebietes müsste die Sprache ST naturgemäss eine Echtzeitsprache sein. Das Normungsgremium scheint jedoch keine der existierenden Echtzeitsprachen zu kennen und hat deshalb das Rad neu erfunden — wobei die nun vorgeschlagenen Echtzeitfähigkeiten dabei eher bescheiden bis unzureichend ausgefallen sind. So sind absolute Zeitangaben und Möglichkeiten zur zeitlichen Überwachung von Aktivitäten nicht vorgesehen, Tasks können allein zyklisch eingeplant werden und schliesslich ist die Verzögerung das einzige vorhandene Mittel zur Steuerung des Zeitverhaltens. Insofern erwächst PEARL durch die bevorstehende Normung von ST keine ernstzunehmende Konkurrenz. Dies gilt auch bezüglich der übrigen — konventionellen — Sprachelemente von ST. Die Brauchbarkeit von ST (und der anderen IEC-Sprachen) wird vielmehr deutlich unter dem Mangel an höheren und implementationsunabhängigen Ein/Ausgabekonzepten und an Möglichkeiten zur Programmierung verteilter Systeme leiden.

Der einzige Punkt, in dem ST den Leistungsumfang von PEARL übertrifft, ist die direkte anwendungsbezogene Formulierbarkeit sequentieller Ablaufsteuerungsaufgaben mit Hilfe der drei Sprachelemente (Abb. 1)

- (Anfangs-) Schritt

- Transition

- Aktion

Damit können an industrielle Steuerungsanwendungen angepasste Petri-Netze, mit denen die Koordinierung und Kooperation asynchroner sequentieller Prozesse beschrieben werden, in textueller Form wiedergegeben werden. Um die führende Rolle von PEARL als *die* Sprache für den Automatisierungstechniker auch weiterhin zu gewährleisten, ist es unbedingt notwendig, Sprachmittel zur Unterstützung sequentieller Ablaufsteuerungen ebenfalls in PEARL zur Verfügung zu stellen.

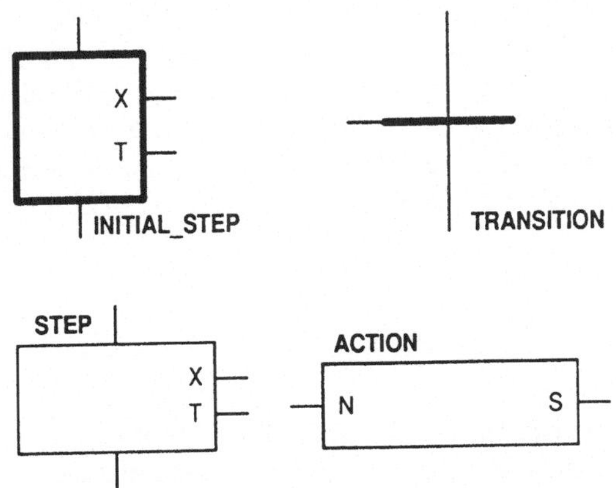

Abbildung 1. Sprachelemente sequentieller Ablaufpläne

Nach der Definition geeigneter neuer Sprachkonstrukte kann dies in zwei Formen geschehen:

1. durch die Erweiterung der existierenden Compiler oder

2. durch die Verwendung eines Präprozessors.

Der zweite Weg wird in einem an der Reichsuniversität Groningen durchgeführten Projekt beschritten, das im Rahmen dieses Artikels ausführlich beschrieben werden soll.

Zuerst geben wir im nächsten Abschnitt eine kurze Zusammenfassung der theoretischen Grundlagen sequentieller Abläufe. Danach werden die zur Formulierung von Ablaufsteuerungen benötigten Sprachmittel derart definiert, dass sie sich problemlos in die bestehende PEARL-Syntax einfügen, und dann in Backus-Naur-Form exakt beschrieben. Auf der Grundlage der Semantik sequentieller Abläufe wird gezeigt, wie sich diese Sprachmittel durch in bisherigem PEARL formulierte Codesequenzen auflösen lassen. So reduzieren sich Schritte i.w. auf Ansprungmarken und Transitionen werden durch bedingte Verzweigungen realisiert. In Abhängigkeit von ihrem Typ werden Aktionen in Codestücke, Prozeduraufrufe bzw. Task-Aktivierungen umgesetzt. Folgen auf einen Schritt mehrere alternative Abläufe, die durch sich gegenseitig ausschliessende Transitionsbedingungen eingeleitet werden, so müssen letztere jeweils alle nach jeder Ausführung des Schrittes ausgewertet werden. Kompliziert ist eigentlich nur die Generierung von Code zur Implementierung paralleler Ablauffolgen. Hierzu ist für jeden parallelen Zweig eine Hilfsprozedur anzulegen, die den Code für jeden in ihrem Zweig vorkommenden Schritt und der nachfolgenden Transition in den Alternativen einer Fallunterscheidung enthält. Die Verzweigung zu den einzelnen Fällen erfolgt durch beim Aufruf der Prozeduren übergebene Hilfsparameter, die von den Prozeduren beim Erfülltsein einer Transitionsbedingung fortgeschaltet werden. Die Transitionsbedingung, welche die parallelen Abläufe wieder in eine einzige Sequenz zusammenführt, wird nur dann abgefragt, wenn sich alle parallelen Zweige in ihrem jeweils letzten Schritt befinden.

Durch das hier dargestellte Präprozessorprojekt wird konstruktiv gezeigt, dass sich geeignete Elemente zur Unterstützung der Formulierung sequentieller Ablaufsteuerungen ohne Schwierigkeiten in den Sprachumfang von PEARL integrieren lassen. Dadurch wird PEARL in die Lage versetzt, auch weiterhin das ganze Spektrum der in der Prozessautomatisierung benötigten Funktionalität bereitzustellen.

2. Sequentielle Ablaufpläne

Das mathematische Modell sequentieller Ablaufpläne zur Darstellung des Geschehens in industriellen Prozessen ist aus der bekannten Petri-Netz-Theorie abgeleitet worden und dient zur übersichtlichen Formulierung der Koordination asynchroner Vorgänge. Eine industrielle Steuerung wird statisch durch einen Ablaufplan dargestellt. Ihre Dynamik erschliesst sich durch die Interpretation des Planes unter Beachtung bestimmter semantischer Regeln. Ein sequentieller Ablaufplan ist ein gerichteter Graph definiert als das Quadrupel

$$(S,T,V,I)$$

wobei

$S = \{s_1, ..., s_m\}$ eine endliche, nicht leere Menge von Schritten,
$T = \{t_1, ..., t_n\}$ eine endliche, nicht leere Menge von Transitionen,
$V = \{v_1, ..., v_k\}$ eine endliche, nicht leere Menge von Verbindungen zwischen jeweils einem Schritt und einer Transition bzw. einer Transition und einem Schritt und

schliesslich

$I \subset S$ die Menge der Anfangsschritte ist.

Die Mengen S und T stellen die Knoten des Graphen dar. Bei der Aufstellung eines solchen Graphen sind die folgenden Regeln zu beachten:

- Es muss mindestens ein Anfangsschritt vorhanden sein.

- Jedem Schritt können Aktionen zugeordnet sein.

- Zu jeder Transition existiert genau eine Transitionsbedingung.

- Die alternierende Folge von Schritten und Transitionen muss unbedingt eingehalten werden, d.h.

    - zwei Schritte dürfen niemals direkt verbunden werden, sondern sie müssen immer durch eine Transition getrennt sein und

    - zwei Transitionen dürfen ebenfalls nie direkt ohne die Trennung durch einen Schritt verbunden werden.

Die Anfangsschritte werden zu Beginn der Ausführung aktiviert und bestimmen so den Anfangszustand. Ein Schritt ist ein logischer Zustand eines Steuerungssystems, während dessen es sich gemäss der Regeln verhält, die in den dem Schritt zugeordneten Aktionen festgelegt sind. Sind mit einem Schritt keine Aktionen assoziiert, so hat der Schritt die Funktion, auf das Erfülltsein der nachgeschalteten Transitionsbedingung zu warten. Zu einem gegebenen Zeitpunkt während der Ausführung eines Systems

- kann ein Schritt entweder aktiv oder inaktiv sein und

- der Zustand der SPS ist durch die Menge der aktiven Schritte bestimmt.

Die einem Schritt zugeordneten Aktionen werden bei seiner Aktivierung mindestens einmal und danach so lange wiederholt ausgeführt, wie sich der Schritt im aktiven Zustand befindet. Während eines Schrittes können Aktionen initiiert, fortgesetzt oder beendet werden. Das Ende eines Schrittes ist durch das Auftreten von Prozessereignissen charakterisiert, auf Grund derer die Bedingung für den Übergang zum nächsten Schritt erfüllt wird. Deshalb können sich Schritte nicht überlappen.

Transitionen stellen Bedingungen zur Steuerung des Kontrollflusses von jeweils einem oder mehreren ihnen vorausgehenden Schritten zu einer Anzahl von Nachfolgerschritten entlang der Kanten in einem gerichteten Graphen dar. Eine Transition ist entweder freigegeben oder nicht. Sie ist nur dann freigegeben, wenn alle mit ihr verbundenen und ihr unmittelbar vorausgehenden Schritte aktiv sind. Eine Transition zwischen zwei Schritten kann nur dann ausgeführt werden, wenn

- sie freigegeben ist und

- die Transitionsbedingung den Wert wahr angenommen hat.

Ist nun eine Transition freigegeben und die zugehörige Boole'sche Bedingung erfüllt, so werden alle der Transition unmittelbar vorausgehenden Schritte deaktiviert und alle ihr über Kanten eines gerichteten Graphen unmittelbar nachfolgenden Schritte gleichzeitig aktiviert.

Die mit den einzelnen Schritten eines Ablaufplans assoziierten Aktionen werden in graphischen Darstellungen des Kontrollflusses aus Gründen der Übersichtlichkeit meistens nicht eingezeichnet. Aktionen werden mit einer Kette aus folgenden Zeichen, die von links nach rechts interpretiert werden, näher qualifiziert. Daraus sind Angaben über die Zeitpunkte und -dauern der Aktionen, die Frequenzen ihrer Ausführung usw. zu entnehmen.

N non-stored, unconditional
R reset
S set / stored
P having pulse form
C conditional
L time limited
D time delayed

Die Grundformen sequentieller Ablaufpläne, ein typisches Beispiel sowie einige Typen von Aktionen sind in den Abbildungen 2 – 6 dargestellt.

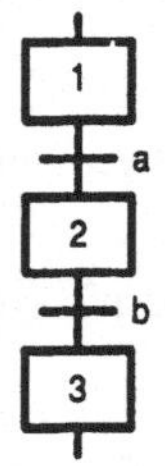

Abbildung 2. Ein linearer Ablaufplan

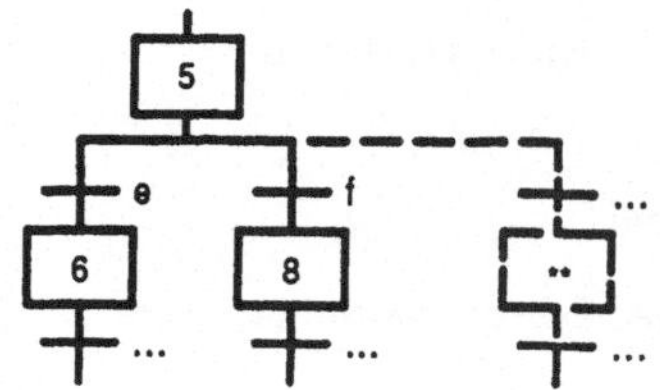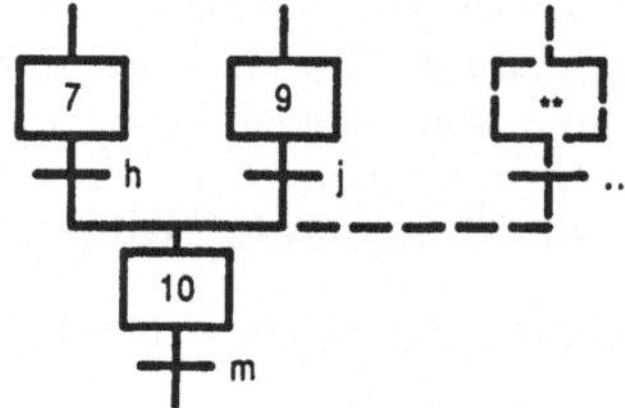

Abbildung 3. Verzweigung und Wiederzusammenführung alternativer Abläufe

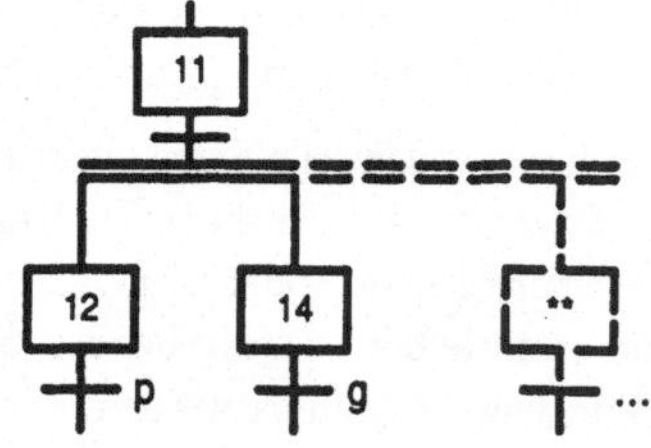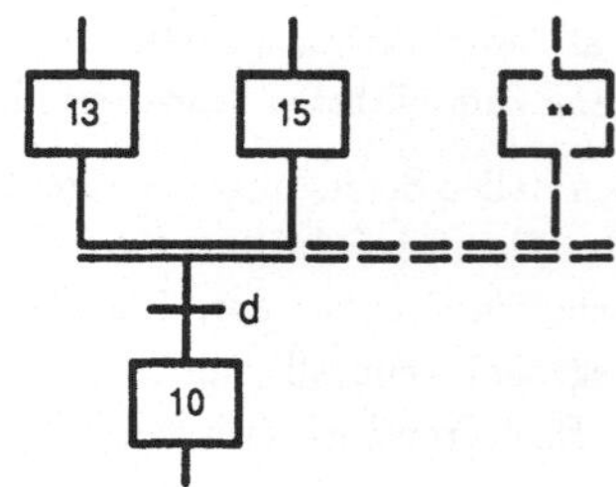

Abbildung 4. Verzweigung und Wiederzusammenführung paralleler Abläufe

3. Formale Definition der PEARL-Erweiterungen

In diesem Abschnitt sollen nun zur Formulierung sequentieller Ablaufpläne geeignete Erweiterungen der PEARL-Syntax definiert werden. Als Gliederungselement für Hauptprogramme, Prozeduren und Tasks fügt sich das hier eingeführte Konzept problemlos in die bestehende PEARL-Syntax gemäss [3] ein. Das wurde durch deutliche Abweichung von der Syntax sequentieller Ablaufpläne nach dem IEC-Entwurf erreicht, ohne jedoch die Funktionalität zu verändern.

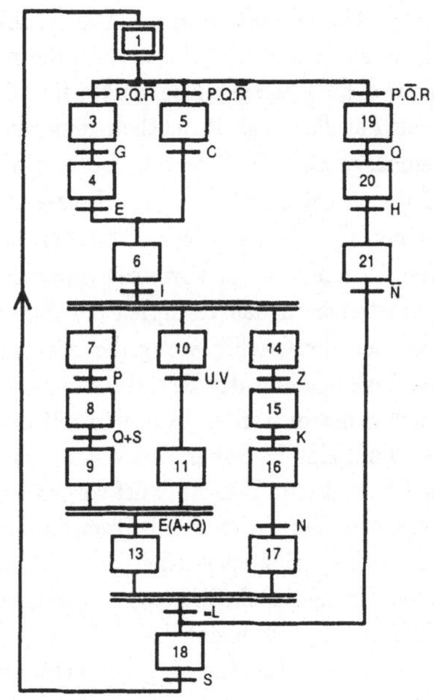

Abbildung 5. Ein typischer sequentieller Ablaufplan

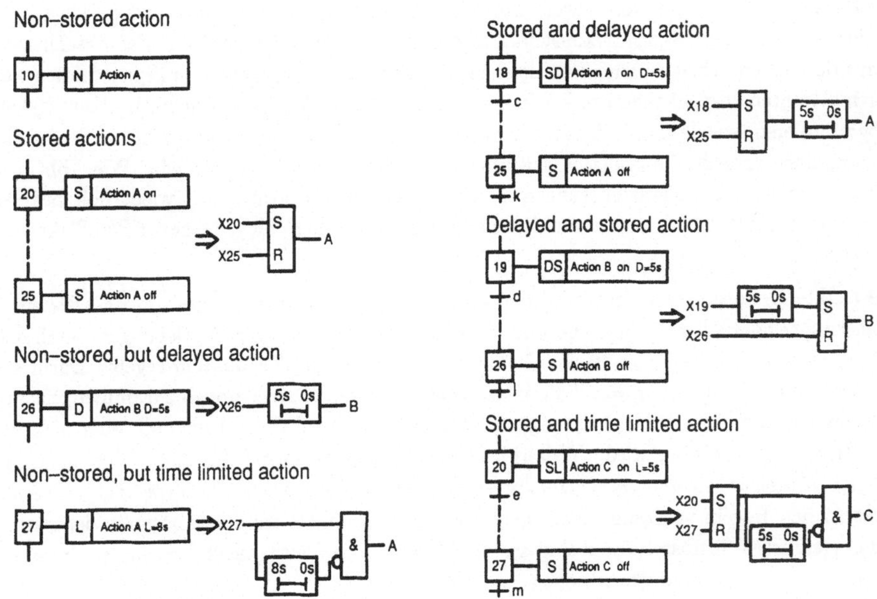

Abbildung 6. Verschiedene Aktionstypen

Im einzelnen wurde folgendermassen von der im IEC-Entwurf beschriebenen (und noch nicht entgültig festgelegten) Syntax zur textuellen Formulierung sequentieller Ablaufpläne abgewichen. Jeder Ablaufplan wird in seiner Gesamtheit durch syntaktische Klammern gekapselt. Auf besonders ausgezeichnete Anfangsschritte konnte verzichtet werden, da die Syntax die Existenz eines ersten Schrittes am Anfang jeden Ablaufplans erzwingt. Die Struktur von Ablaufplänen bzw. der entsprechenden gerichteten Graphen wird durch Schachtelung der neuen Sprachelemente wiedergegeben und durch explizite Verwendung eigener Konstrukte für die Auswahl von Ablaufalternativen und zur Formulierung von Paralleltätigkeit deutlich gemacht. Dadurch erübrigt sich die im IEC-Entwurf vorgesehene Namensgebung für Schritte. Dort werden auch Aktionen mit Namen versehen, um sie mit deren Hilfe Schritten zuzuordnen. Dieses sowie die Möglichkeit, Aktionen explizit zu deklarieren, erscheinen überflüssig. Deshalb werden in der hier vorgestellten Syntax die einem Schritt zugeordneten Aktionen in der Form sequentiellen Codes, von Prozeduraufrufen und Tasking-Anweisungen unmittelbar in den Rümpfen der Schritte aufgeführt und so mit ihnen assoziiert. Ebenso erweist sich die Qualifizierung der Aktionen als unnötig, da in PEARL umfassende Möglichkeiten des Taskings und der Zeitsteuerung zur Verfügung stehen, von denen in den Schrittrümpfen Gebrauch gemacht werden kann und mit denen die gleichen Effekte erzielt werden können. Schliesslich wird nur eine Form der Verzweigung in alternativ zu wählende Ablauffolgen bereitgestellt, denn der Effekt der im IEC-Entwurf vorgesehenen Möglichkeit expliziter Priorisierung der einzelnen Alternativen kann auch durch Umordnung in der Aufschreibung erreicht werden. Als Option kann in [2] noch spezifiziert werden, dass sich die einzelnen, die verschiedenen alternativen Zweige einleitenden, Transitionsbedingungen gegenseitig ausschliessen sollen, so wie es von der Norm IEC 848 gefordert wird. Es bleibt die Aufgabe des Programmierers, diese Eigenschaft der Bedingungen sicherzustellen. Um eine mögliche Fehlerquelle zu vermeiden, wurde auf diese Option in unserem Entwurf ebenfalls verzichtet.

Beim Entwurf der neuen Sprachkonstrukte wurde zur Vermeidung inkonsistenten Verhaltens darauf geachtet, dass Ablaufsteuerungen immer mit einem Schritt enden und dass die einzelnen Zweige des Parallelkonstruktes sowohl jeweils mit einem Schritt beginnen als auch enden. Dagegen sind in den Zweigen der Alternativkonstruktion nach der einleitenden und vor der abschliessenden Transition neben Schritten auch Parallelelemente zugelassen. Die genannten Eigenschaften werden durch die unten angegebenen Produktionsregeln erzwungen. Auf die Möglichkeit, Ablaufzyklen zu programmieren, wurde aus Gründen der Sicherheit bewusst verzichtet und weil die Semantik von Ablaufplänen ohnehin die zyklische Ausführung der einzelnen Schritte vorsieht. Wie Abb. 5 zeigt, werden sequentielle Ablaufsteuerungen in ihrer Gesamtheit jedoch häufig wiederholt bearbeitet. Dies lässt sich in PEARL einfach durch Einbettung eines Ablaufplans in eine REPEAT-Anweisung erreichen.

Die exakte Definition der vorgeschlagenen zusätzlichen Sprachkonstrukte für PEARL zur Formulierung sequentieller Ablaufsteuerungen erfolgt nun gemäss obiger Überlegungen durch die folgenden sieben, in erweiterter Backus-Naur-Form angegebenen, Produktionsregeln. Darin stehen das Zeichen | zur Auflistung von Wahlmöglichkeiten, die mit () zu Gruppen zusammengefasst werden können, und [] zur Kennzeichnung von Optionen, während mit { } die null- oder mehrmalige Wiederholung bezeichnet wird. Weiterhin stehen 'statement-string' für eine Folge ausführbarer PEARL-Anweisungen und 'Boolean-expression' für einen Ausdruck des Typs BIT(1). Alle anderen in den Regeln vorkommenden nicht terminalen Symbole werden dort auch definiert. Die neu eingeführten Schlüsselwörter sind selbsterklärend. Die Syntaxregeln erzwingen alternierende Folgen von Schritten und Transitionen.

```
sfc::= SEQUENCE sfc-body ENDSEQ;

sfc-body::= step [(transition | alternatives) (sfc-body | parallel)]

alt-body::= (step | parallel) [(transition | alternatives) alt-body]

step::= STEP [statement-string] ENDSTEP;
```

```
transition::= TRANSITION Boolean-expression;


parallel::= PARALLEL THREAD sfc-body {THREAD sfc-body} ENDPAR;


alternatives::= SELECT BRANCH transition alt-body transition
                      {BRANCH transition alt-body transition} ENDSCT;
```


4. Arbeitsweise des Präprozessors

Auf der Grundlage der im zweiten Abschnitt beschriebenen Semantik sequentieller Ablaufpläne soll nun ausgeführt werden, wie unser Präprozessor die oben definierten neuen Sprachkonstrukte in Codestücke umwandelt, die in Standard-PEARL formuliert sind.

Schritte reduzieren sich auf Ansprungmarken am Beginn von Anweisungsfolgen, welche die den Schritten zugeordneten Aktionen realisieren. Wie bereits erläutert, werden in unserer Spracherweiterung Aktionen in Abhängigkeit von ihrem Typ durch Codesequenzen, Prozeduraufrufe oder Tasking-Operationen wiedergegeben. Transitionen werden in bedingte Verzweigungen abgebildet. Somit werden einfache sequentielle Abläufe, wie sie z.B. in Abb. 2 dargestellt sind, nach folgendem Muster in PEARL-Programmstücke umgeformt:


```
step: sequentieller Code;
      Prozeduraufrufe;
      Tasking-Anweisungen;
      ;
      etc. gemaess der Typen der einzelnen Aktionen
      .

      .

      .

      IF NOT transitions-bedingung THEN GOTO step FIN;
      .

      .

      .
```


Folgen auf einen Schritt mehrere alternative Abläufe (vgl. Abb. 3), die durch eigene Transitionsbedingungen eingeleitet werden, so müssen letztere jeweils alle nach jeder Ausführung des Schrittes ausgewertet werden, sowie es im folgenden Programmstück angedeutet ist:


```
step_0: sequentieller Code;
        .

        .

        .

        IF transitions-bedingung_1 THEN GOTO step_1 FIN;
        IF transitions-bedingung_2 THEN GOTO step_2 FIN;
        .

        .

        .

        IF transitions-bedingung_N THEN GOTO step_N ELSE GOTO step_0 FIN;

step_i: sequentieller Code;
```

```
        .
        .
    IF transitions-bedingung_i1 THEN
        GOTO Naechster Schritt in i-ter Alternative ELSE GOTO step_i FIN;
        .
        .
        .
```

Entsprechend der in [2] angegebenen Semantik stellt obige Auflösung alternativer Ablaufsequenzen sicher, dass bei gleichzeitigem Erfülltsein mehrerer Transitionsbedingungen, die jeweils eine Alternative einleiten, genau eine ausgewählt wird, und zwar die zuerst aufgeschriebene bzw. die in graphischer Darstellung am weitesten links erscheinende. Schliessen sich die Transitionsbedingungen gegenseitig aus, so führt die oben beschriebene Auflösung natürlich zum Ziel. Ansonsten werden Mehrdeutigkeiten durch Verzweigung in die zuerst notierte erfüllte Alternative vermieden.

Nicht so offensichtlich und weniger übersichtlich als in den letzten beiden Fällen gestaltet sich die Generierung von Standard-PEARL-Code zur Implementierung paralleler Ablauffolgen (vgl. Abb. 4). Es liegt nahe, hierfür die Echtzeitfähigkeiten von PEARL auszunutzen, indem für jeden der parallel auszuführenden Zweige eine Hilfs-Tasks angelegt wird. Beim Erfülltsein der dem Parallelkonstrukt vorausgehenden Transitionsbedingung wären alle diese Hilfs-Tasks mit gleicher Priorität sofort zu aktivieren und die Task, welche die sequentielle Ablaufsteuerung umfasst, wäre bis zur Beendigung aller parallelen Zweige zu suspendieren. Von dieser Möglichkeit wurde jedoch Abstand genommen, da sich Schwierigkeiten beim Wiederzusammenführen der einzelnen Zweige ergeben und weil ausserdem hierzu ein Sprachmittel notwendig wäre, das in PEARL nicht vorhanden ist. Darüberhinaus soll in unserer Implementierung unabhängig von Laufzeit- und Betriebssystem sichergestellt werden, dass parallele Zweige immer derart sequentialisiert werden, dass in der Reihenfolge der Aufschreibung von jedem Zweig jeweils ein Schritt einmal bearbeitet und die dann nachfolgende Transitionsbedingung abgefragt werden, bevor danach im Nachbarzweig in gleicher Weise fortgefahren wird. Um diese Eigenschaft zu realisieren, wird für jeden parallelen Zweig eine Hilfsprozedur angelegt. Der Rumpf einer solchen Prozedur besteht aus einer Fallunterscheidung, in der jedem in dem entsprechenden Zweig vorkommenden Schritt zusammen mit der nachfolgenden Transition eine Alternative zugeordnet ist. Die Verzweigung zu den einzelnen Fällen erfolgt durch beim Aufruf der Prozeduren übergebene Steuerparameter, welche von den einzelnen Prozeduren beim Erfülltsein einer Transitionsbedingung fortgeschaltet werden. Die allen Zweigen gemeinsame Transitionsbedingung, die jeweils dem letzten Schritt der einzelnen Zweige folgt und die parallelen Abläufe wieder in eine einzige Sequenz zusammenführt, wird nur dann abgefragt, wenn sich alle parallelen Zweige in ihrem jeweils letzten Schritt befinden. Als PEARL-Programmstück sieht die so beschriebene Konstruktion folgendermassen aus:

```
    DECLARE l BIT, index FIXED(10);
        .
        .
        .
step_0: sequentieller code;
        .
        .
        .
    IF NOT transitions-bedingung_0 THEN GOTO step_0 FIN;
    index(1):=1; ... ; index(Anzahl-Zweige):=1;
par:    l:='1'B;
    l:=l AND proc_1(index(1));
```

```
        .
        .
        .
     l:=l AND proc_Anzahl-Zweige(index(Anzahl-Zweige));
     IF NOT l THEN GOTO par FIN;
     IF NOT transitions-bedingung_F THEN GOTO par FIN;

step_F: . . .
        .
        .
```

mit den folgendermassen definierten N Hilfsprozeduren

```
proc_i: PROCEDURE (ind FIXED) RETURNS (BIT);
     DECLARE r BIT;
     r:='0'B;
     CASE ind
        ALT sequentieller Code des ersten Schrittes;
            IF erste-transitions-bedingung THEN ind:=ind+1 FIN;
        ALT analog fuer den zweiten Schritt
        .
        .
        .
        ALT sequentieller Code des letzten Schrittes; r:='1'B
     FIN;
     RETURN(r);
     END;
```

Bei der oben umrissenen Umsetzung sequentieller Ablaufsteuerungen in PEARL-Programmstücke
wurde stillschweigend vorausgesetzt, dass die verschiedenen Zweige der Alternativ- und Parallelkon-
strukte aus einfachen linearen Schritt-Transitions-Abfolgen bestehen. Im allgemeinen werden die
neuen Sprachkonstrukte jedoch auch ineinandergeschachtelt, um komplexere Ablaufstrukturen zu
formulieren. Entsprechend sind bei ihrer Transformation in Standard-PEARL auch die nach obigen
Mustern gebildeten Programmsequenzen zu verschachteln.

Literatur

[1] W. A. Halang: "Schwerpunkte der internationalen Forschung im Bereich Echtzeitsysteme".
    In: R. Henn, K. Stieger (Hrsg.), PEARL 89 — Workshop über Realzeitsysteme, 10. Fachta-
    gung des PEARL-Vereins e.V., Boppard, Dezember 1989. Informatik-Fachberichte 231, pp.
    1 – 12. Berlin-Heidelberg-New York: Springer-Verlag 1989.

[2] International Electrotechnical Commission, Technical Committee 65: Industrial Process Mea-
    surement and Control, Subcommittee 65A: System Considerations, Working Group 6: Dis-
    continuous Process Control, Working Draft "Standards for Programmable Controllers", Part
    3: "Programming Languages", IEC 65A(Secretariat)90-I, Dezember 1988.

[3] DIN 66 253: Programmiersprache PEARL, Teil 1 Basic PEARL, 1981; Teil 2 Full PEARL,
    1982. Berlin: Beuth-Verlag.

Software-Entwurf für Automatisierungssysteme mit PEARL

G. Thiele
Institut für Automatisierungstechnik, FB 1, Universität Bremen
Kufsteiner Straße, 2800 Bremen 33

Zusammenfassung

Zielsetzung und Eignung von PEARL sollten es insbesondere dem Automatisierungsingenieur nahe legen, den Software-Entwurf für Realzeit-Systeme "PEARL-orientiert" durchzuführen. Ein derartiges "Entwurfs-Denken in PEARL" kann einmal dadurch gefördert werden, daß die Stärken von PEARL an konkreten Automatisierungs-Problemen verdeutlicht werden. Auf der anderen Seite muß erkennbar sein, daß Lösungen auf niedrigerer Abstraktionsebene PEARL-orientiert überzeugend dargestellt werden können und Lösungen auf abstrakterer Ebene in PEARL darstellbar und leichter zu verstehen sind.

Zu den Darstellungsmitteln gehören neben den Erweiterungen der Ausdrucksmittel von Struktogrammen auf Module und Tasks im Sinne von PEARL auch die Einführung von Modul-Bäumen als Import-Graphen und von auf Tasks verallgemeinerten Funktions-Strukturbäumen. Konzeptionell von grundlegender Bedeutung ist das funktionelle Verständnis der Realzeitverwaltung im Sinne eines von PEARL "mitgelieferten" Realzeitverwaltungs-Moduls.

Die genannten Darstellungsmittel werden gegenüber früheren Veröffentlichungen in erweiterter Form vorgestellt und diskutiert. Außerdem werden an Hand von veröffentlichten Problemlösungen, wie z.B. der Programmierung von konkurrierenden Tasks für Steuerungsprobleme in einer eigens hierfür geschaffenen Pascal-ähnlichen Sprache [2] , die Möglichkeiten eines PEARL-orientierten Entwurfs verdeutlicht.

1. Einleitung

Das PEARL-Task-Modell ist begrifflich und funktionell, auch von der verwendeten Abstraktions-Ebene her, auf die Aufgaben des Automatisierungs-Ingenieurs zugeschnitten. Durch die Möglichkeit des expliziten Einplanens von Tasks zeichnet sich PEARL gegenüber den meisten anderen Realzeit-Sprachen aus, in denen z.B. eine zeitzyklische Task-Ausführungs-Einplanung auf niedrigerem Niveau und damit mit größerem Anwender-Aufwand bei geringerer Transparenz formuliert werden muß. In Abschnitt 2 werden verschiedene Varianten in PEARL-orientierter Form gegenübergestellt.

Das "Denken in PEARL" legt, anders als das abstraktere Denken in Daten- und Kontroll-Transformationen [10] ,das Einbeziehen von Tasks bereits beim Software-Entwurf nahe. Es zeigt sich, daß mit entsprechenden Erweiterungen, wie sie im PEARL 90 - Sprachvorschlag [1] formuliert sind, ein PEARL-orientierter Entwurf von Steuerungsproblemen, im Sinne einer für solche Problemstellungen entwickelten Pascal-ähnlichen Sprache [2], problemlos möglich ist (Abschnitt 3).

Die Ergänzung von Entwurfs-Konstrukten, die den Entwurf erleichtern, müssen nicht von vornherein vom "Denken in PEARL" zum Denken in einer anderen Sprache führen, jedenfalls dann nicht, wenn eine systematische Umsetzung des erweiterten Entwurfs auch in PEARL möglich ist. In Abschnitt 4 wird dies am Beispiel des in [3] vorgestellten Software-Entwurfs für eine konfigurierbare Feldstation im Sinne einer Objekt-orientierten Formulierung der Funktionsblöcke gezeigt.

2. Entwurfsmittel auf PEARL-Ebene

2.1 Verallgemeinerter Funktions-Strukturbaum

Die expliziten Task-Anweisungen von PEARL erlauben eine Verallgemeinerung des Funktionsstruktur-Baums als Strukturierungshilfsmittel auf Tasks mit Task-Anweisungs-Bezügen [4]. Dabei können diese (z.B. zeitzyklische Einplanung oder Einplanung auf Interrupts oder Taskbeendigung), gegebenenfalls in abgekürzter Form (z.B. für die unmittelbare Ausführungs-Einplanung bzw. die Ausführungseinplanung ohne Einplanungsvorsatz), sowie Taskprioritäten auch explizit angegeben werden (Bild 1).

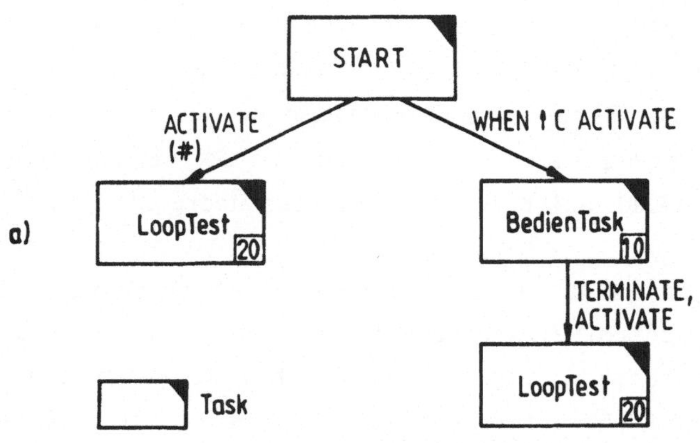

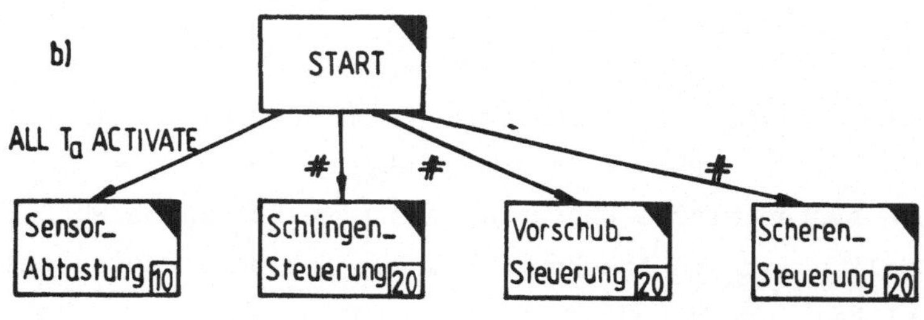

Bild 1: Funktionsstrukturbaum für den "Loop-Test" der analogen
Prozeß-E/A (a) und für eine Steuerung (b).

2.2 Import-Graph

Der Import-Graph [5] als Strukturierungshilfsmittel beim Entwurf
im Großen wird zweckmäßigerweise ebenfalls in Baum-Struktur
entworfen, da er dann bezüglich der importierten Prozeduren und
Tasks als Aggregation eines korrespondierenden

Funktionsstrukturbaums aufgefaßt werden kann (Bild 2). Die

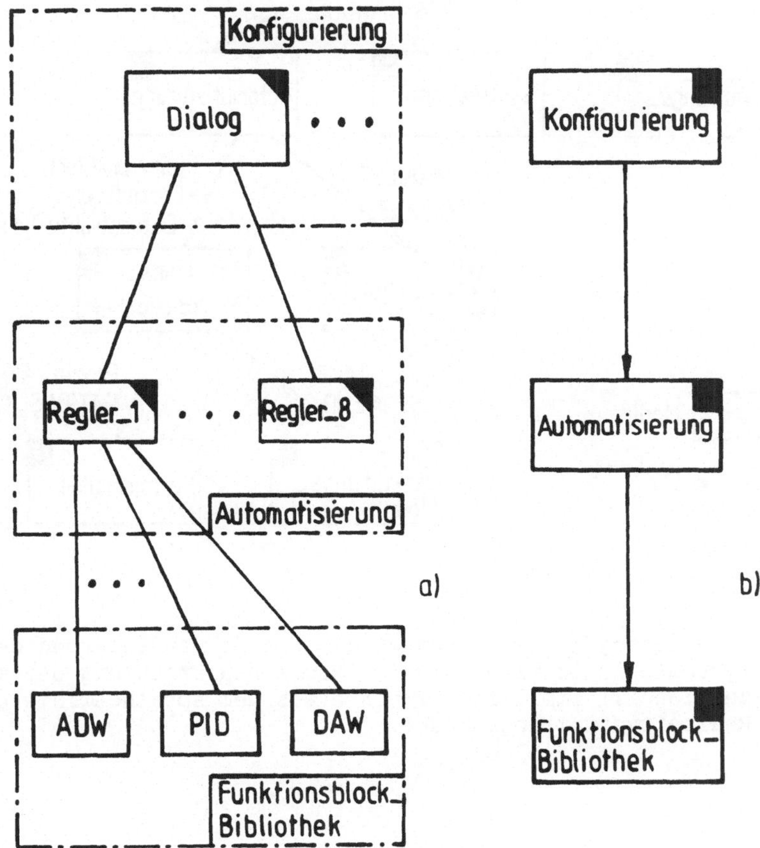

Bild 2: Importgraph in Form eines Modulbaums (b) und korres-
        pondierender Funktionsstrukturbaum (a) für den Soft-
        wareentwurf einer konfigurierbaren Feldstation (Legen-
        de S. Bild 1 bzw. Bild 3).

Baumstruktur kann auch bezüglich der in PEARL im Systemteil
deklarierten Anwendergerätenamen durchgehalten werden, wenn der
Systemteil in einen eigenen Modul ausgelagert wird [4] . Die
Importe aus dem "sprachdefiniert mitgelieferten" Realzeit-
Verwaltungsmodul von PEARL sind als implizite Importe aufzufassen
(Bild 3).

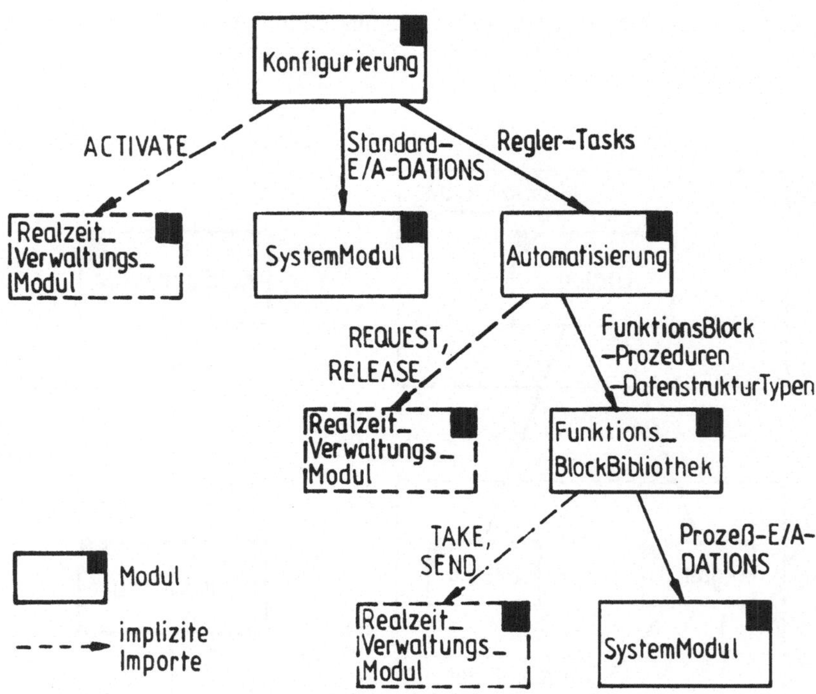

Bild 3: Erweiterung des Modul-Strukturbaums bei Auslagerung des
Systemteils in einen eigenen Modul und Ergänzung der (kon-
zeptionell) impliziten Importe aus dem sprachdefinierten
Realzeitverwaltungs-Modul.

2.3 Task-Kopplungs-Graph

Die Taskkopplungen, z.B. durch Daten-Austausch über Modul-lokale
Variablen oder durch gegenseitige Synchronisation, sind im Rahmen
eines Funktionsstrukturbaums i.a. nicht darstellbar. Als
Entwurfshilfsmittel ist hierfür ein Task-Kopplungs-Graph bzw.
Task-Struktur-Graph [6] geeignet. In Bild 4 ist ein solcher

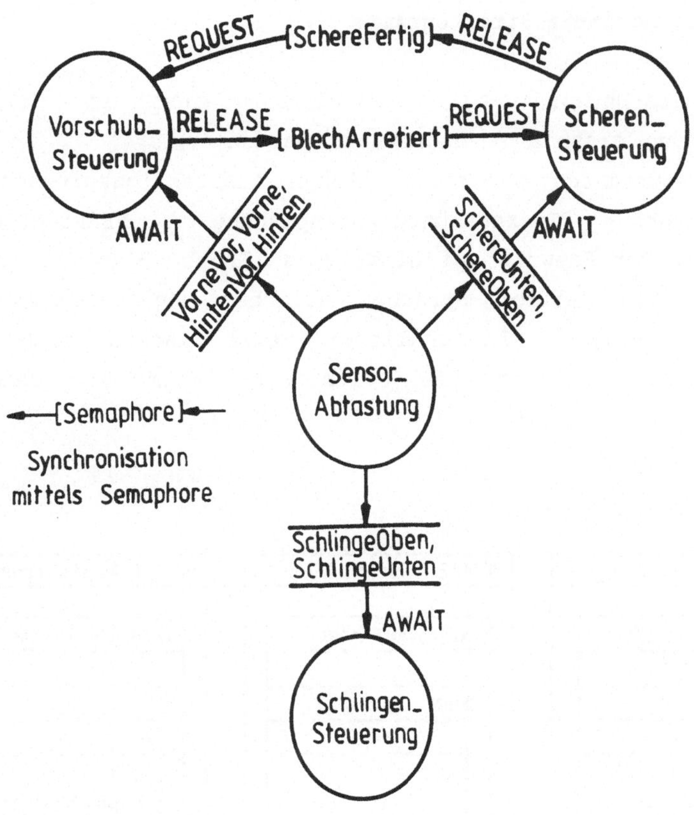

Bild 4: Task-Kopplungs-Graph für eine Steuerung.

Graph für das Beispiel der Steuerung einer Bandquerteilanlage [2] angegeben. Die Tasks "VorschubSteuerung" und "ScherenSteuerung" synchronisieren sich wechselseitig über die Semaphoren "SchereFertig" und "Blech Arretiert"("enge" Kopplung [6]). Alle Steuerungstasks sind wiederum über Bedingungsvariablen mit der zeitzyklisch eingeplanten Task "SensorAbtastung" "lose" gekoppelt, wobei AWAIT "Aktives Warten" der entsprechenden Task auf das Erfülltsein einer entsprechenden Bedingung bedeutet.

2.4 PEARL-orientierte Struktogramme

Der Software-Entwurf auf PEARL-Ebene kann in Form von Struktogrammen graphisch unterstützt werden [13,4] . In Bild 4 sind zwei Varianten unterschiedlichen Abstraktionsniveaus einer zeitzyklischen Taskausführungseinplanung PEARL-orientiert dargestellt. Der Entwurf auf niedrigerer Ebene (Bild 5b) gibt die in den meisten Realzeit-Sprachen üblichen Möglichkeiten hierzu wieder [6,7] . In ähnlicher Weise lassen sich andere Entwurfsformen [8] in PEARL-orientierter Form veranschaulichen (Bild 5c).

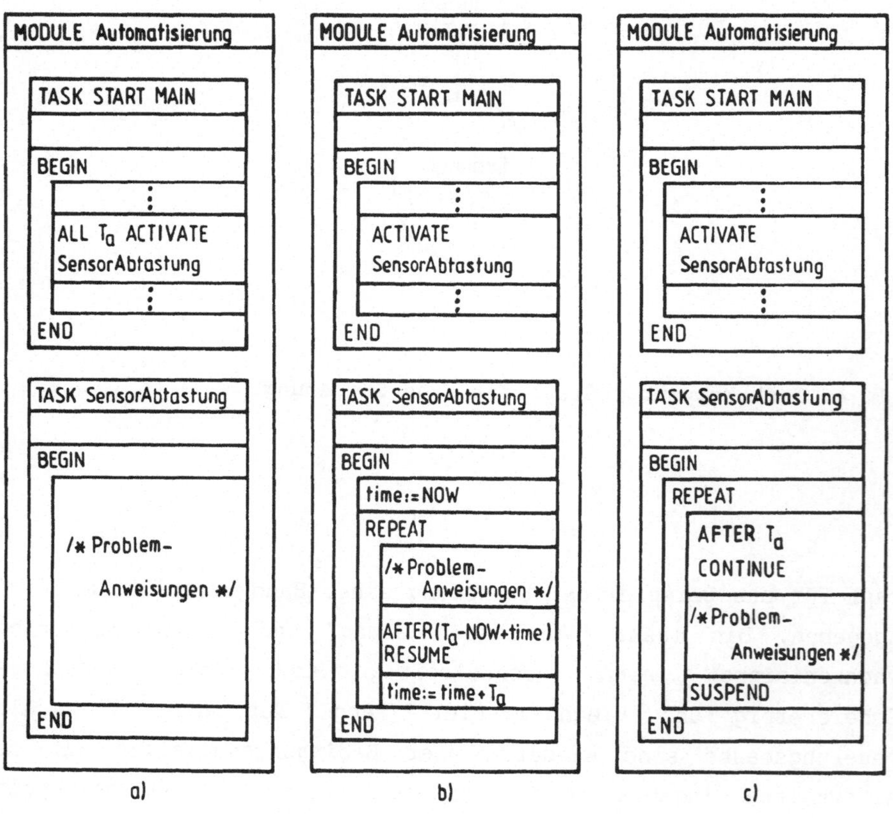

Bild 5: Zeitzyklische Task-Ausführungs-Einplanung auf höherem (a) und niedrigerem (b,c) Abstraktions-Niveau.

3. PEARL-orientierter Entwurf einer Steuerung

Charakteristisch für Steuerungsabläufe ist u.a. das "Aktive Warten" auf das Eintreten von Ereignissen oder auf Meldungen, wenn unmittelbar darauf eine Aktion ausgelöst werden muß, z.B. das Abschalten des Antriebs eines Anlagenteils. Werden verschiedene derartige Anlagenteile parallel gesteuert, um die Leistungsfähigkeit der Anlage zu erhöhen [2] , so kann es sein, daß mehrere Tasks gleichzeitig auf unterschiedliche Ergeignisse aktiv warten müssen. Durch Round-Robin-Einplanung von mit gleicher Priorität deklarierten Tasks, wie dies auch im PEARL 90 - Sprachvorschlag [1] vorgesehen ist, kann vermieden werden, daß das Aktive Warten einer Steuerungstask (in MSRBASIC [12] Sequenz genannt) das Erkennen eines anderen Ereigniseintritts durch eine konkurrierende Steuerungstask verhindert. PEARL besitzt zwar kein "AWAIT"-Sprachkonstrukt, wie es in [2,12] eingeführt ist, es kann aber äquivalent durch eine entsprechende wiedereintrittsfähige Prozedur gleichen Namens ersetzt werden (Bild 6).

Geht das gesteuerte Anlagenteil ohne Abschaltung in eine Ruhelage über, so genügt eine hinreichend lange Task-Verzögerung mit der daran anschließenden Ausführung des nächsten Steuerschritts.

Das zeitzyklische Abfragen der Meßsensoren kann eine entsprechend eingeplante Sensor-Abtastungstask , die mit höherer Priorität als die der Steuerungstasks deklariert ist,übernehmen unter der Voraussetzung, daß der Meßvorgang die Reaktionszeit der Steuerungstasks auf aktiv erwartete Ereignisse nur unwesentlich vergrößert.

Task-Verzögerung und Aktives Warten sind Möglichkeiten für den Steuerungsentwurf mittels konkurrierend oder parallel ausführbarer Task-Abschnitte. Hängt jedoch die Ausführung des nächsten Steuerschrittes eines Anlagenteils (z.B. Vorschub) vom Erreichen eines bestimmten Zustandes eines anderen Anlagenteils (z.B. Schere) ab, so bietet sich hier die Synchronisation der beiden Tasks mittels Semaphoren an.

In Bild 6 ist dargestellt, wie die Steuerungssoftware für eine Bandquerteilanlage [2] PEARL-orientiert entworfen werden kann.

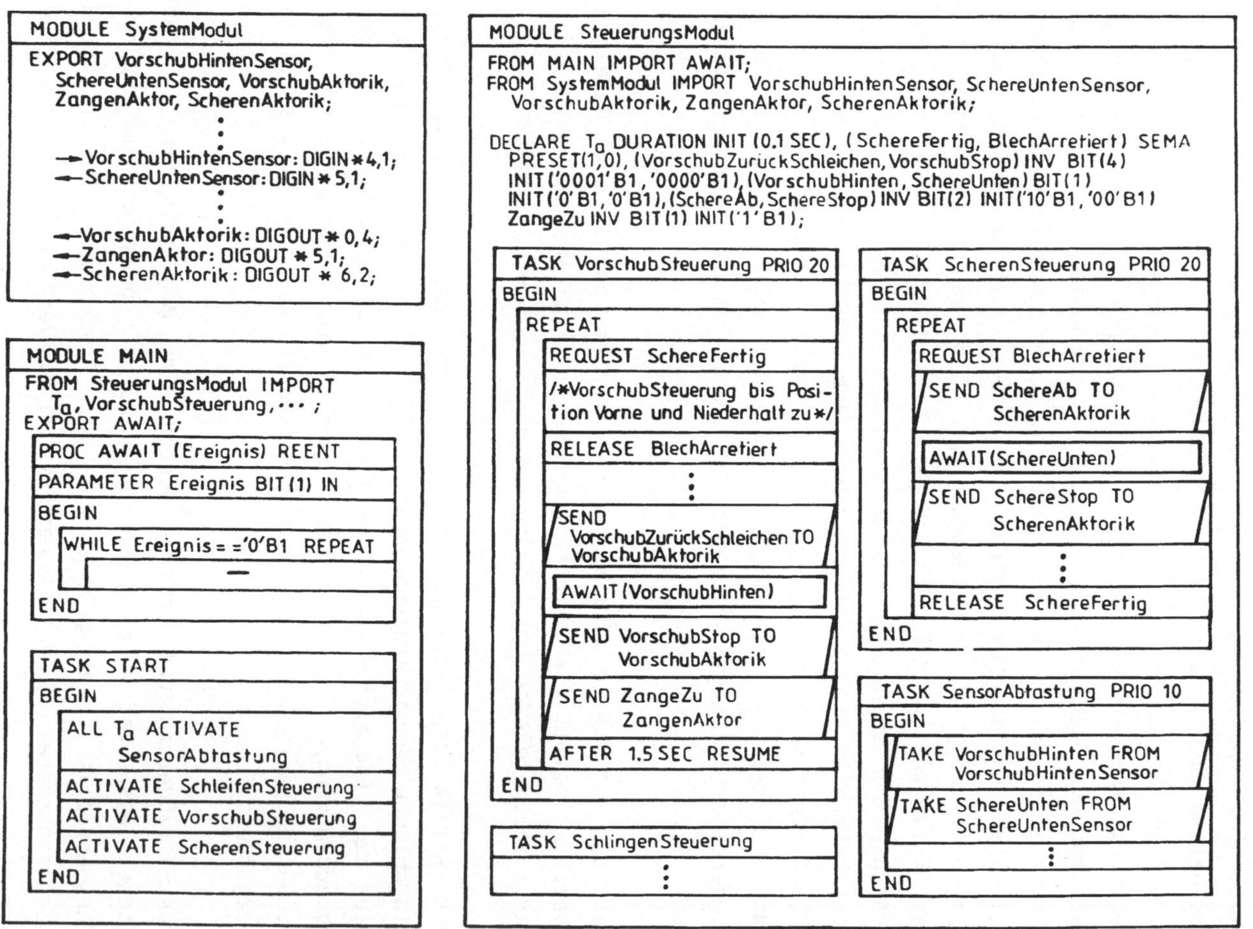

Bild 6: PEARL-orientierter Entwurf einer Steuerung.

4. Erweiterung der PEARL-orientierten Entwurfsmittel im Sinne von Objektklassen

Typischerweise beeinhaltet der Funktionsblock-Bibliotheksmodul der Software einer konfigurierbaren Feldstation die Funktionsblock-Prozeduren und die Typen der Funktionsblock-Datenstrukturen [3].

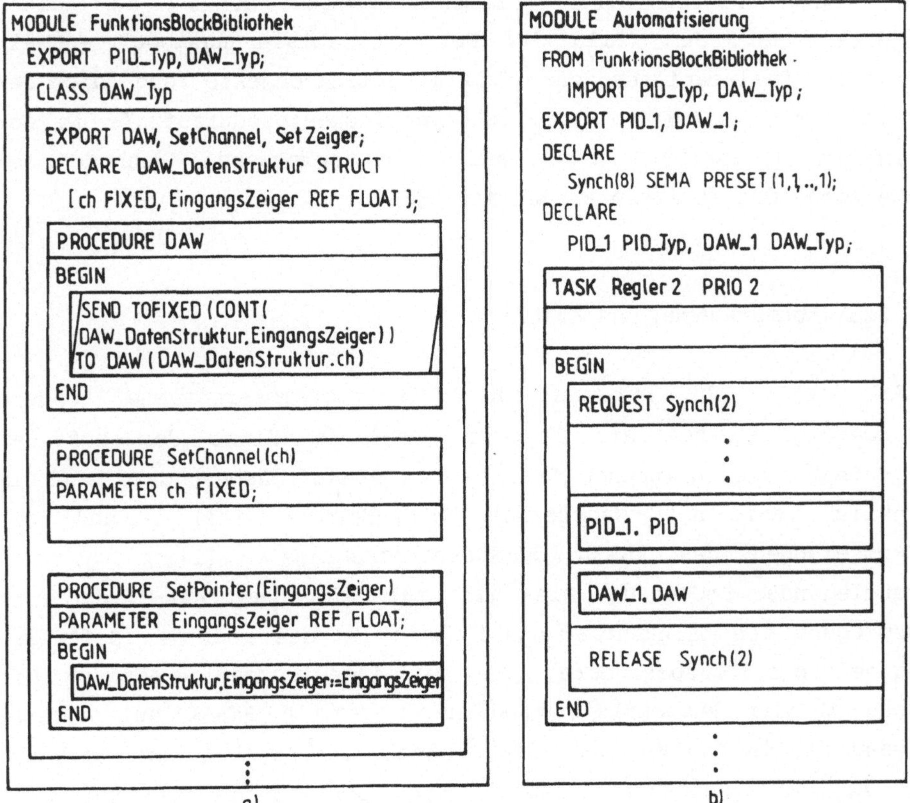

Bild 7: Entwurf des Funktionsblock-Bibliotheks-Moduls in Objekt-orientierter Form (a) und zugehöriger Entwurf des Auto-matisierungsmoduls (b) einer konfigurierbaren Feldstation in PEARL-orientierter Ausprägung.

Strukturell bietet sich eine Zusammenfassung des jeweiligen Datenstrukturtyps und der Prozedur zum Typ eines Objekts höheren Abstraktionsgrades, das heißt zu einer Objekt-Klasse, an. Dies soll hier in Anlehnung an [9] dargestellt werden.

Die eigentliche Datenstruktur eines im Automatisierungsmodul deklarierten Funktionsblock-Objektes seiner Klasse kann hierbei außerhalb des Objektes unsichtbar bleiben und die Konfigurierung der Verbindungen und die Parametrierung mittels entsprechender Zugriffs-Prozeduren der Objektklasse durchgeführt werden.
Der Funktionsblock-Bibliotheksmodul wird somit zu einem Objektklassen-Modul (Bild 7). Der Automatisierungsmodul enthält nun die Deklarationen der Funktionsblock-Objekte und die der Regler-Tasks, in denen die Funktionsblock-Prozeduraufrufe in der Form qualifizierter Zugriffe auf die entsprechende "Methode" der jeweiligen Objektklasse erscheinen (Bild 7).

5. Schlußbemerkungen und Ausblick

PEARL ist nicht nur als Realzeit-Programmiersprache sondern darüberhinaus auch als Entwurfsmittel im Sinne des auf den Ingenieur zugeschnittenen "Denkens in PEARL",insbesondere auch im Hinblick auf die Mehrrechner-PEARL-Norm [11] und die Erweiterungen des PEARL 90-Sprachvorschlags [1] , von grundlegender Bedeutung. Dies gilt auch für die PEARL-orientierte Ausprägung von Strukturierungshilfsmitteln des modernen Software-Engineerings. Insbesondere z.B. der ingenieurmäßige Umgang mit Tasks in der Automatisierungstechnik ist in PEARL auf höherem Niveau als in den meisten anderen Realzeit-Sprachen bzw. Sprachen mit Realzeiterweiterungen möglich und deshalb für den Software-Entwurf vorteilhaft einzusetzen, gegebenenfalls auch in erweiterter Form. Dies gilt auch dann, wenn die Implementation in einer anderen Sprache erfolgt, da die Implementation dann als systematische Umsetzung des PEARL-orientierten Entwurfs durchführbar und wartbar ist.

6. Literatur

[1]   Stieger, K. (1989). PEARL 90 - Die Weiterentwicklung von
        PEARL. In R. Henn, K. Stieger (Hrsg.): PEARL 89 - Workshop
        über Realzeitsysteme, Informatik-Fachberichte 231, Springer-
        Verlag, Berlin, pp. 99-136.

[2]   Ley, F. und M. von Wüllen (1989). Eine freiprogrammierbare
        Steuerung zur Realisierung von parallelen Ablaufstrukturen.
        Automatisierungstechnische Praxis atp, 30, H.7., pp.343-349.

[3]   Thiele, G., D. Popovic, U. Claussen und E. Wendland (1990).
        Entwurf und Implementierung einer Menü-geführten graphisch
        konfigurierbaren und Multi-Tasking-fähigen Mehrregelkreis-
        Feldstation. In H.Rzehak, L. Drebinger (Hrsg.): Proc. Echt-
        zeit 90, Sindelfingen 1990, pp. 147-156.

[4]   Thiele, G., D. Popovic, P. Baacke, P. Flügel und L. Renner
        (1990). Strukturierter Entwurf, Implementation und Valida-
        tion eines Menü-geführten Identifikations-Programmpakets auf
        einer PEARL-Engine. In R. Henn, K. Stieger (Hrsg.): PEARL
        89 - Workshop über Realzeitsysteme, Informatik-Fachberichte
        231, Springer-Verlag, Berlin, pp. 196-209.

[5]   Blaschek, G., G. Pomberger und F. Ritzinger (1987). Einfüh-
        rung in die Programmiersprache Modula - 2. Springer-Verlag,
        Berlin.

[6]   Nielsen, K. und K. Shumate (1988). Designing Large Real-Time
        Systems with Ada. McGraw-Hill Book Company, New York.

[7]   Inmos Ltd. (1988). Communicating Process Architecture.
        Prentice-Hall, New York, ch. 8, pp. 105-120.

[8]   Künzel, O. (1990). Meßdatenverarbeitung einer Meteo-Station
        mit Pascal-RTK. In H. Rzehak, L. Drebinger (Hrsg.): Proc.
        Echtzeit 90, Sindelfingen 1990, pp. 449-456.

[9]   Meyer, B. (1988). Object-oriented Software Construction.
        Prentice-Hall, New York.

[10]   Mellor, S.J. und P.T. Ward (1986). Structured Development
        for Real-Time Systems. Vol.3, Yourdon Press, Englewood
        Cliffs.

[11]   DIN 66253, Teil 3 (1989). Programmiersprache PEARL: Mehr-
        rechner-PEARL. Beuth-Verlag, Berlin.

[12]   Schmidt, G. (1985). MSRBASIC-Kurzübersicht. Lehrstuhl für
        Steuerungs- und Regelungstechnik, Universität München.

[13]   Thiele, G. (1987). Strukturierter Entwurf von Realzeit-Al-
        gorithmen für Mikrorechner in der Prozeßautomatisierung.
        Berichte Elektrotechnik, Nr. 4/87, Universität Bremen.

Eine Übersicht über die Software-Entwurfsmethode HOOD

T. Tempelmeier, München

Zusammenfassung: Die Software-Entwurfsmethode HOOD der European Space Agency (ESA) wird in ihren wesentlichen Aspekten dargestellt. Dabei werden einerseits die zugrunde gelegten Konzepte und Repräsentationsformen und andererseits die planmäßige Vorgehensweise von HOOD erläutert. Ein Vergleich mit anderen Methoden zeigt bezüglich der Konzepte und Repräsentationsformen nur geringe Unterschiede zu HOOD. Hinsichtlich der Vorgehensweisen ergeben sich jedoch starke Unterschiede, die aber in der Praxis zu einer einheitlichen Entwurfsrichtlinie integriert werden können. Abschließend ist ein erster Vorschlag angegeben, wie ein in HOOD erstellter Entwurf nach PEARL umgesetzt werden könnte.

Stichwörter: HOOD, OOD, Ada, PEARL, (Hierarchical) Object Oriented Design, Software-Entwurf, Software-Engineering, Luft- und Raumfahrt.

1. Einleitung

Die Software-Entwurfsmethode HOOD (Hierarchical Object Oriented Design) wurde seit 1986 im Auftrag der European Space Agency (ESA) entwickelt. Ihr Einsatz wird für die geplanten Raumfahrtprojekte der ESA und für weitere Großprojekte im Luftfahrtbereich verbindlich vorgeschrieben. HOOD wird über den Bereich der Großunternehmen hinaus auch alle Zulieferbetriebe mit Software-Anteilen betreffen. Insofern ist eine rechtzeitige Auseinandersetzung mit HOOD dringend geboten.

In der von 1987 bis 1989 gültigen Version 2.2 hat HOOD bereits eine gewisse Bekanntheit erlangt. Mit der seit Ende 1989 verbindlichen Version 3.0 [1,2] ergaben sich jedoch einige erhebliche Änderungen. Hier wird nur die aktuelle Version 3.0 behandelt.

HOOD lehnt sich stark an die Programmiersprache Ada an. Einige Konzepte von Ada wurden aber in HOOD durch andere ersetzt oder es wurde HOOD gegenüber Ada erweitert. Während Version 2.2 konzeptionell noch sehr ähnlich zu Ada war, ergeben sich mit Version 3.0 stärkere Abweichungen.

Anschrift des Autors: Dr. T. Tempelmeier, Elfenstraße 39, D - 8000 München 83, Telefon 089 / 602 622, Email: tempelmeier @ lrz-extern.uni-muenchen.dbp.de. Der Autor ist Professor für Informatik an der Fachhochschule Rosenheim und befaßte sich früher bei der Firma MBB und seither im Rahmen verschiedener Beratungstätigkeiten mit HOOD und ähnlichen Vorgehensweisen.

HOOD wurde primär für die Phase des Software-Grobentwurfs entwickelt. Unter dem Begriff Grobentwurf, oder auch Software-Architektur, versteht man den Aufbau eines Software-Systems aus seinen grundlegenden Bausteinen; diese Bausteine sind in der Regel die Tasks und Module des Systems (vgl. etwa [3]).

Darüberhinaus soll HOOD auch den Übergang zum Feinentwurf und einen Teil der Feinentwurfsphase abdecken.

Im folgenden werden zunächst die Konzepte und dann die planmäßige Vorgehensweise in HOOD beschrieben. Die praktische Anwendbarkeit von HOOD, ein Vergleich mit verwandten Methoden und erste Überlegungen zum Übergang von HOOD nach PEARL beschließen den Beitrag.

2. Konzepte und ihre grafischen / textuellen Repräsentationsformen

Als Grundbausteine für den Software-Entwurf stellt HOOD *aktive und passive Objekte* zur Verfügung. Diese entsprechen in erster Näherung den Tasks und Paketen in Ada. Die grafische Repräsentation von Objekten ist aus Bild 1 ersichtlich. Objekte werden durch Rechtecke mit abgerundeten Ecken dargestellt, wobei aktive Objekte

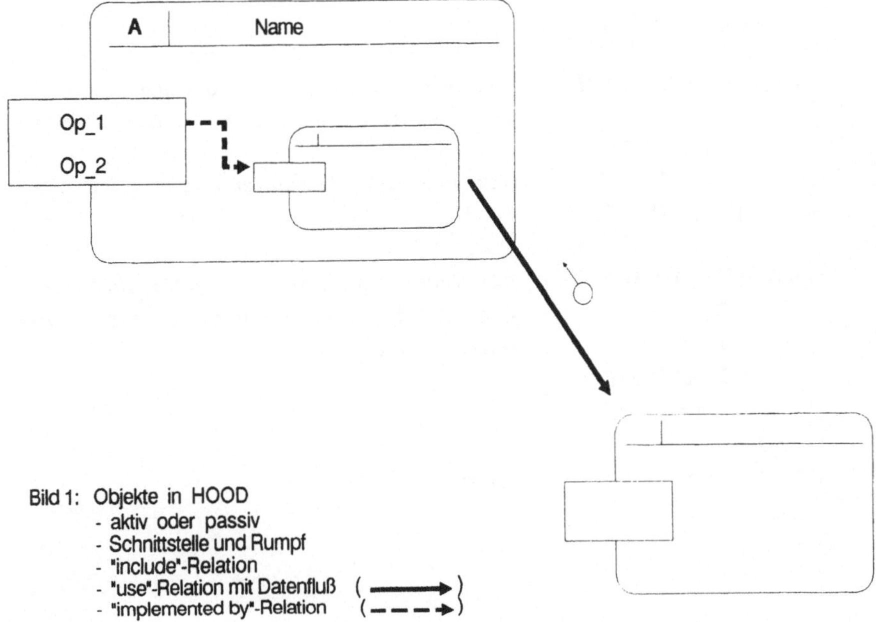

Bild 1: Objekte in HOOD
- aktiv oder passiv
- Schnittstelle und Rumpf
- "include"-Relation
- "use"-Relation mit Datenfluß (———→)
- "implemented by"-Relation (— — — →)

zusätzlich den Buchstaben A in der linken oberen Ecke tragen. Die Schnittstelle der Objekte nach außen, in HOOD *Provided Interface* genannt, wird durch ein vorstehendes Rechteck symbolisiert, in welches die exportierten Operationen des Objekts eingetragen werden.

Objekte können andere Objekte enthalten; HOOD spricht diesbezüglich von einer *Parent-Child-* oder *Include-Relation.* Diese Beziehung entspricht der von Ada und anderen Programmiersprachen bekannten Verschachtelung. Falls Operationen der Schnittstelle direkt durch ein enthaltenes Objekt realisiert werden, so definiert HOOD dies als *Implemented_by-Relation.*

Werden durch ein Objekt die Ressourcen benützt, die ein anderes, nicht im ersten Objekt enthaltenes Objekt zur Verfügung stellt, so spricht HOOD von der *Use-Relation.* Diese Beziehung entspricht der üblichen Import-Beziehung (in Ada: "with").

Zu jedem Objekt gibt es als textuelle Beschreibung das sogenannte *Object Description Skeleton (ODS),* welches alle Informationen zum Objekt in einer Ada-ähnlichen Syntax noch einmal detailliert enthält. Durch dieses ODS kann eine Überfrachtung der grafischen Darstellungen vermieden werden. Das ODS ist wie folgt aufgebaut:

```
OBJECT name IS   active_or_passive

    DESCRIPTION
      -- informeller Text

    IMPLEMENTATION_OR_SYNCHRONISATION_CONSTRAINTS
      -- informeller Text

    PROVIDED_INTERFACE    -- Ada-Syntax; entspricht der Paketschnittstelle;
        TYPES             -- exportierte Typen, Konstanten und Unterprogramme.
        CONSTANTS
        OPERATIONS        -- Unterprogramme (Funktionen und Prozeduren).
        EXCEPTIONS

    REQUIRED_INTERFACE    -- Ada-Syntax; entspricht der Importbeziehung (in
        OBJECTS           -- Ada: "with"), jedoch werden die benötigten Elemente
        TYPES             -- einzeln aufgelistet.
        OPERATIONS
        EXCEPTIONS

    DATAFLOWS             -- informeller Text
```

```
OBJECT_CONTROL_STRUCTURE
            -- Beschreibt das Aufrufverhalten von aktiven Objekten.
            -- Basiert auf dem Ada-Rendezvous-Konzept, ergänzt
            -- um informellen Text.


INTERNALS            -- Ada-Syntax; entspricht dem Paketrumpf.
    OBJECTS
    DECLARATIONS
    OPERATIONS       -- Unterprogramme (Funktionen und Prozeduren).


OPERATION_CONTROL_STRUCTURE
            -- Ada-Pseudocode aller Funktionen und Operationen.
            -- Entspricht den Ada-Unterprogrammrümpfen.
```

END OBJECT name ;

Von besonderer Wichtigkeit ist das OBCS *(Object_Control_Structure)*. Mit diesem wird das Synchronisations- und Kommunikationsverhalten der aktiven Objekte, d.h. im wesentlichen der Tasks, beschrieben. Das OBCS wird mit Ada-Rendezvous-Konstrukten, die um zusätzlichen informellen Text erweitert werden müssen, angegeben. Für andere Zielsprachen als Ada erlaubt HOOD ausdrücklich, ..."andere Implementierungssprachen mit den zugehörigen Betriebssystemprimitiven" im OBCS anzugeben ([1], Kapitel 10.3).

Mit Hilfe des ODS besteht in eingeschränktem Maße auch die Möglichkeit, generische Pakete (HOOD-Bezeichnung: *Class*) zu definieren. Nur für die instantiierten Objekte sind hierbei grafische Darstellungen vorgesehen.

Als *Virtual_Node* - Objekte (Kennbuchstabe V in der linken oberen Ecke der grafischen Darstellung) werden Zusammenfassungen von Objekten bezeichnet, die gewisse Bedingungen erfüllen und dadurch als Verteilungseinheiten in Mehrrechner- oder Mehrprozessorsystemen geeignet sind.

Schließlich können noch *Environment* - Objekte definiert werden (Kennbuchstabe E), die den Bezug auf eine vorgegebene Schnittstelle für das zu entwerfende System erlauben. Man kann sich dies als eine Ada-Paketspezifikation ohne Paketrumpf vorstellen.

3. Anleitung für ein planmäßiges Vorgehen

Der Begriff "Methode" beinhaltet unauflöslich auch Regeln für ein planmäßiges Vorgehen oder eine Verfahrensweise. In HOOD wird in diesem Zusammenhang eine hierarchische Top-Down-Zerlegung der Objekte vorgeschlagen (vgl. Bild 2):

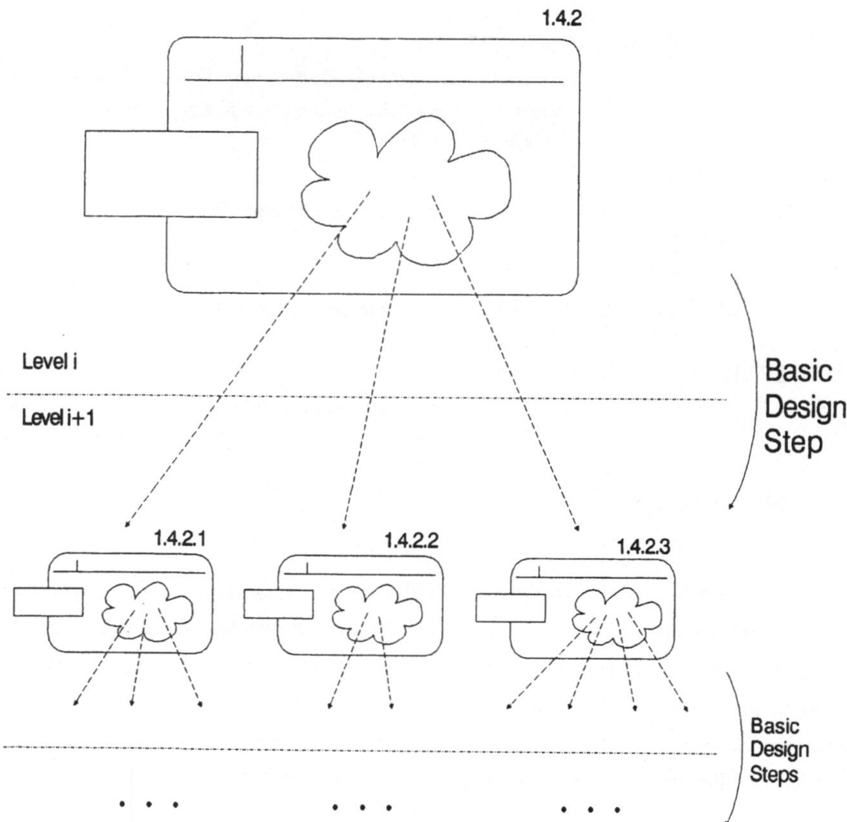

Bild 2: Der "Basic Design Step" in HOOD und der
daraus resultierende "Design Process Tree"

• Der gesamte Entwurfsvorgang wird in eine Anzahl von einzelnen Entwurfsschritten
(*Basic Design Steps*) aufgeliedert. In einem *Basic Design Step* wird ein Objekt
entsprechend der *Include*-Relation in seine Kind-Objekte zerlegt. Daraus ergibt
sich der *Design Process Tree*, der genau der Hierarchie der *Include*-Relation entspricht.

• Die einzelnen Entwurfsschritte (*Basic Design Steps*) werden wiederum in Einzel-
aktivitäten aufgegliedert. Es handelt sich dabei um eine Weiterentwicklung der Ideen
von BOOCH [4,5]. Im einzelnen ist folgender Ablauf für einen *Basic Design Step*
vorgesehen:

> 1. Problem Definition
>   1.1 Statement of the Problem
>   1.2 Analysis and Structuring of the Requirement Data

2. Elaboration of an Informal Solution Strategy
3. Formalisation of the Strategy
   3.1 Identification of Objects
   3.2 Identification of Operations
   3.3 Grouping Operations and Objects
   3.4 Graphical Description
   3.5 Justification of the Design Decisions
4. Formalisation of the Solution - Object Description Skeleton

Die vollständige Beschreibung eines bestimmten Entwurfsobjekts wird sukzessive in zwei *Basic Design Steps* entwickelt:

- In der Entwurfsebene i werden die benötigten Kind-Objekte identifiziert und in einem unvollständigen ODS mit ihrer Schnittstellenspezifikation abgelegt (*"Was muß das Objekt leisten?"*).

- In einem späteren *Basic Design Step* in der Entwurfsebene i+1 wird dann das Objekt selbst als Vater-Objekt betrachtet und durch Spezifikation seiner Interna vervollständigt (*"Wie erbringt das Objekt seine Leistung?"*).

Alle Objekte sind entsprechend dem *Design Process Tree* hierarchisch durchnumeriert. Beispielsweise bezeichnet die Objektnummer 1.4.2 das zweite Kind des Objekts 1.4 usw. Für jedes (!) Objekt ist eine Dokumentation des Entwurfsvorgangs entsprechend der Gliederung eines *Basic Design Steps* anzufertigen. Das entsprechende Gliederungsschema wird auch als *HOOD Chapter Skeleton (HCS)* bezeichnet. Durch Kombination der Objektnumerierung mit der HCS-Kapitelnumerierung ist auch die Struktur der Gesamtdokumentation vorgegeben.

4. Gütekriterien und Entwurfsleitlinien

Prinzipien für die Objektbildung sind klar (wenn auch zu knapp) als Bestandteil der Methode HOOD ausgewiesen. Leider sind für einige schwierige Fragen, z.B. nach welchen Gesichtpunkten aktive Objekte entstehen sollen (dies entspricht in etwa dem Taskentwurf) oder wie die Einhaltung von Zeitbedingungen erreicht werden soll, keine Leitlinien enthalten.

Bezüglich einer Güte-Beurteilung der erstellten Entwürfe stützt sich HOOD auf allgemein anerkannte Kriterien wie hohe Bindung, geringe Kopplung, großes Fan-in, kleines Fan-out, usw. ab.

Insgesamt wird HOOD seinem Anspruch, eine <u>Methode</u> zu definieren, gerecht. HOOD geht über eine bloße Repräsentationsform für Software-Entwürfe hinaus und bietet auch eine systematische Verfahrensweise sowie Gütekriterien an.

5. Praktischer Einsatz von HOOD

Gegenwärtig sind dem Autor drei brauchbare Toolsysteme, die die Konzepte, Repräsentationsformen und die Vorgehensweise von HOOD unterstützen, bekannt. Die Toolsysteme sind auf VMS- und Unix-Workstations ablauffähig. Eines dieser Toolsysteme wird direkt durch die Firma Digital Equipment Corporation vermarktet.

Es gibt zur Zeit nur einige wenige Projekterfahrungen mit HOOD, die öffentlich zugänglich sind [6,7]. Diese beziehen sich in der Regel auf ältere HOOD-Versionen und einen Einsatz ohne ausreichende Tool-Unterstützung. Es sei nicht verschwiegen, daß auch kritische Stimmen zu HOOD laut wurden [8].

6. Bezug zu ähnlichen Methoden

Es gibt eine Reihe von Methoden, die ebenfalls für die Programmiersprache Ada geeignet sind, insbesondere:
• OOD oder Object-Oriented Development [4,5],
• die Methode von Buhr [9],
• OOSD oder Object-Oriented Structured Design [10] sowie
• die Methode nach Nielsen / Shumate [11].

Ein Vergleich dieser Methoden mit HOOD zeigt zunächst, daß hinsichtlich der zugrunde gelegten Konzepte und der angebotenen Repräsentationsformen nur geringe Unterschiede bestehen. Die augenfälligen Ähnlichkeiten der grafischen Darstellungsformen dieser Methoden untereinander und mit HOOD sind in Bild 3 dargestellt.

Einige neuere Arbeiten [12, 13, 14] bestätigen durch ihre ähnlichen bzw. übereinstimmenden Konzepte die durch HOOD vorgegebene Linie.

Ein Methodenvergleich sollte statt der Repräsentationsformen aber eher die angebotenen planmäßigen Vorgehensweisen vergleichen. Hier ergeben sich stärkere Unterschiede: OOSD erhebt schon vom Titel der Erstveröffentlichung her gar nicht den Anspruch, eine bestimmte Vorgehensweise vorzugeben. OOD muß als weniger detaillierter Vorgänger von HOOD angesehen werden. Bei Buhr überwiegt ebenfalls die Betonung der Konzepte und Repräsentationsformen.

Nielsen / Shumate konzentrieren sich fast ausschließlich auf eine systematische, praxisorientierte Verfahrensweise; das Fehlen eines werbewirksamen Methodennamens und ansprechender Grafikelemente dürften die einzigen "Nachteile" dieser sehr fundierten Methode sein.

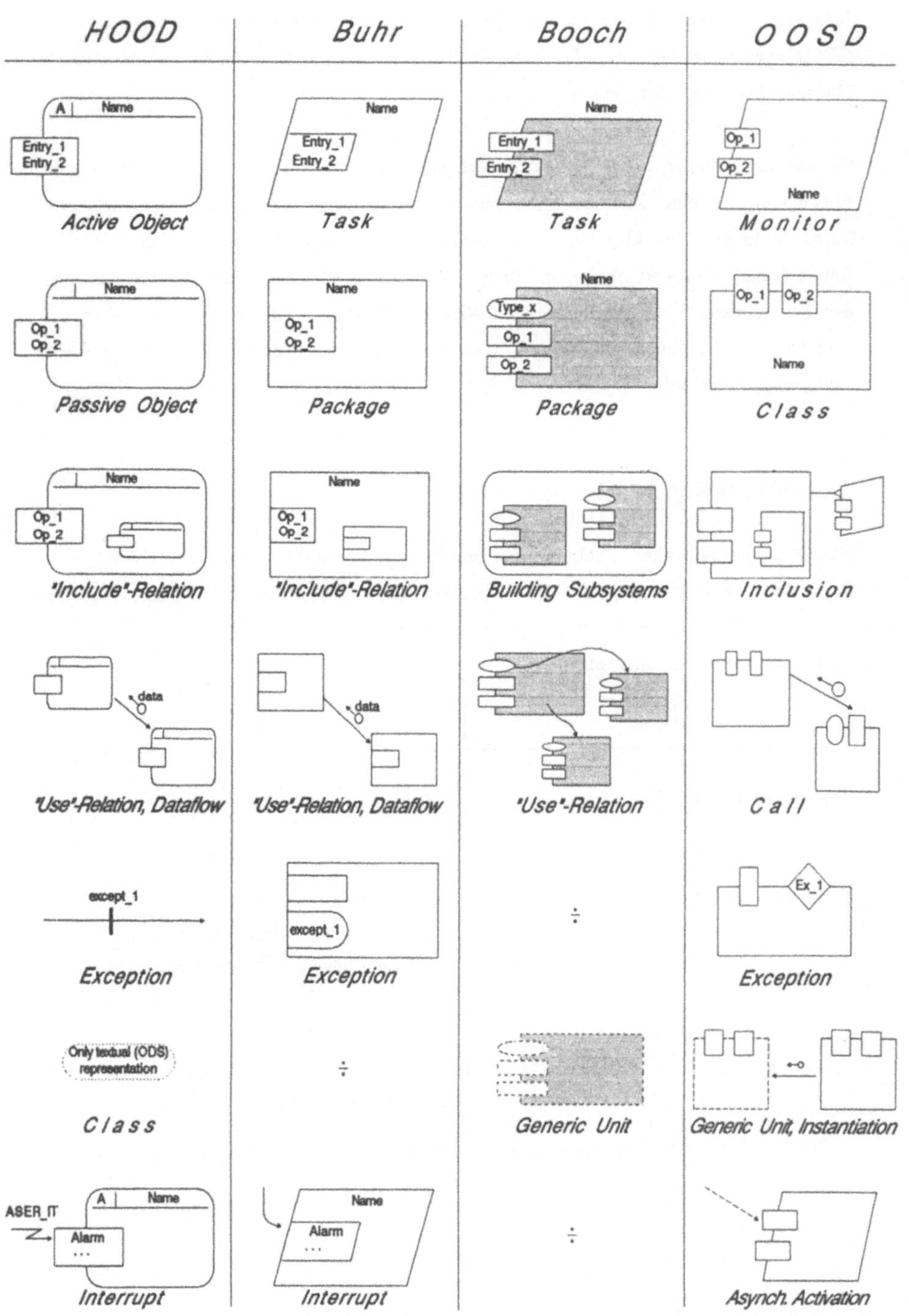

Bild 3: Vergleich der Konzepte und Grafikelemente der Methoden HOOD, Buhr, Booch, OOSD

Insgesamt ist nach Meinung des Autors eine Trennung in Repräsentationsformen (z.B. OOSD oder Buhr) und in umfassende Anleitungen für ein planmäßiges Vorgehen (z.B. Nielsen / Shumate) der richtige Ansatz.

Gerade für weniger erfahrene Entwerfer ist eine systematische Führung durch den Entwurfsprozeß eine wichtige Hilfe, auch wenn sie kein Ersatz für eine fundierte Software-Engineering-Ausbildung sein kann. HOOD ist damit den meisten anderen Entwurfsmethoden weit überlegen. Andererseits ist HOOD noch weit vom Niveau und der Praxisrelevanz der Methode nach Nielsen / Shumate entfernt. Es steht aber jedem frei, die Leitlinien von Nielsen / Shumate anzuwenden oder deren Ideen in Form anwendungsspezifischer Richtlinien in HOOD einzubringen.

7. Abbildung von HOOD nach PEARL

HOOD ist grundsätzlich nicht auf Ada beschränkt. Vielmehr soll es mit HOOD auch für andere Sprachen möglich sein, objektorientierte Entwürfe zu erstellen [1]. Im folgenden werden in tabellarischer Form erste Überlegungen zu einem Übergang von HOOD nach PEARL angegeben:

HOOD	PEARL
Passives Objekt	Modul
Aktives Objekt	Task oder Menge von Tasks
Include – Relation	nach Kenntnis des Autors nicht direkt abbildbar; ersatzweise Verwendung der Use-Relation; zur Fragwürdigkeit der streng hierarchischen Top-Down-Zerlegung in HOOD siehe auch [8]
Use – Relation	Verwendung globaler Objekte mittels SPC; ev. Namenskonventionen nötig
Class – Objekt mit Instantiierungen	Präprozessor mit Macro-Textexpansion

Virtual Node – Objekt
    mit Operationen
    TRANSMIT, RECEIVE, ...
	COLLECTION

Environment – Objekt
    mit Operationen eines PEARL-
    Betriebssystems
	PEARL-Betriebssystemdienste

Object Control Structure – OBCS	Angabe von PEARL-Synchronisations- und Tasking-Anweisungen im OBCS

Literaturangaben

[1]   HOOD Working Group: *HOOD Reference Manual. Issue 3.0.*
European Space Agency, Noordwijk, The Netherlands, Sept. 1989.

[2]   HOOD Working Group: *HOOD User Manual. Issue 3.0.*
European Space Agency, Noordwijk, The Netherlands, Dec. 1989.

[3]   Tempelmeier, T.: *Praxis des Entwurfs von Echtzeit-Softwaresystemen.*
atp Automatierungstechnische Praxis, 31 (1989) 11, 532-538.

[4]   Booch, G.: *Software Engineering with Ada.*
Benjamin/Cummings, Menlo Park, Ca., 1986.

[5]   Booch, G.: *Object-Oriented Development.*
IEEE Transactions on Software Engineering, SE-12 (1986), 211-221.

[6]   Lai, M.: *Why not combine HOOD and Ada? An overview of several French Navy Projects.* In: B. Lynch (ed.): Ada: Experiences and prospects. Proceedings of the Ada-Europe International Conference, Dublin, 12-14 June 1990. Cambridge University Press, Cambridge 1990.

[7]   Mala, W., Grein, C.: *Objektorientierter Entwurf und Programmierung von Echtzeitsystemen.* Fachtagung Prozeßrechensysteme '91, Berlin, 25.-27. Februar 1991. Der Tagungsband wird im Springer-Verlag, Berlin, erscheinen.

[8]   Tempelmeier, T.: *Eine kritische Bewertung der Software-Entwurfsmethode HOOD.* Fachtagung Prozeßrechensysteme '91, Berlin, 25.-27. Februar 1991. Der Tagungsband wird im Springer-Verlag, Berlin, erscheinen.

[9]   Buhr, R.J.A.: *System Design with Ada.*
Prentice-Hall, Englewood Cliffs, N.J., 1984.

[10] Wasserman, A.I., Pircher, P.A., Muller, R.J.: *The Object-Oriented Structured Design Notation for Software Design Representation.* Computer, 23 (1990) 3, 50-63.

[11] Nielsen, K., Shumate, K.: *Designing Large Real-Time Systems with Ada.*
Intertext Publications & McGraw-Hill, New York, N.Y., 1988.

[12] Jalote, P.: *Functional Refinement and Nested Objects for Object-Oriented Design.* IEEE Transactions on Software Engineering, SE-15 (1989), 264-270.

[13] Seidewitz, E.: *General Object-Oriented Software Development: Background and Experience.* Journal of Systems and Software 9 (1989), 95-108.

[14] Wolf, L.W., Clarke, L.A., Wileden, J.C.: *The AdaPIC Tool Set: Supporting Interface Control and Analysis Throughout the Software Development Process.* IEEE Transactions on Software Engineering, SE-15 (1989), 250-263.

Modellierung und Validierung von Exception-Handling-Mechanismen für Realzeitsysteme

Andreas Oberweis

Institut für Angewandte Informatik und Formale Beschreibungsverfahren
Kollegium am Schloß, Bau IV
Universität Karlsruhe
7500 KARLSRUHE
Telefon: (0721)608-4283

Zusammenfassung

Die Zuverlässigkeit von Systemen zur Steuerung, Überwachung und Automatisierung von technischen Systemen hängt wesentlich davon ab, ob ausreichende Mechanismen zur Behandlung von Ausnahmen (Exceptions), d.h. irregulären Ereignissen oder Situationen, vorgesehen sind. Ausnahmen können aber auch als spezielles Programmstrukturierungskonzept eingesetzt werden, um selten auftretende Sonderfälle getrennt vom normalen Systemverhalten beschreiben zu können.

In dieser Arbeit werden Konzepte beschrieben zur Einbeziehung der Ausnahmebehandlung (Exception-Handling) in den Entwurf von Realzeitsystemen. Als Beschreibungssprache werden höhere Petri-Netze (Prädikate/Transitionen-Netze) verwendet. Es werden Möglichkeiten zu einer simulationsgestützten Validierung von Exception-Handling-Mechanismen vorgestellt. Diese Konzepte sind in einen existierenden Petri-Netz-Simulator integriert worden.

1   Einleitung

Ausnahmen (*Exceptions*), d.h. irreguläre Ereignisse oder Situationen, und ihre Behandlung (*Exception Handling*) werden in fast allen Bereichen der praktischen Informatik untersucht [Obe90b], etwa im Zusammenhang mit

- Datenbank- und (Büro-) Informationssystemen,

- Programmiersprachen,

- Realzeitsystemen,

- Robotik,

- Wissensbasierten Systemen,

- Betriebssystemen,

- Fehlertoleranten Systemen.

Bei der Steuerung, Überwachung und Automatisierung von technischen Anlagen ist die Behandlung von Ausnahmesituationen besonders wichtig. Ein unkontrollierter Programmabbruch in einer Ausnahmesituation ist nicht akzeptabel und kann schwerwiegende Folgen haben, beispielsweise daß eine technische Anlage ohne Überwachung und Steuerung durch den Rechner weiterarbeitet und möglicherweise ihre Umwelt gefährdet [HH89].

Entsprechend sind in vielen Programmiersprachen spezielle Sprachmittel zur Behandlung von Ausnahmesituationen eingeführt worden, z.B. in Ada, Clue, Eiffel, Mesa, Pearl (für vergleichende Übersichten siehe [HH89,Fed90,PA90]). Exception-Handling bildet außerdem einen Schwerpunkt bei der Weiterentwicklung von Pearl [Sti89].

In der Entwurfsphase von Realzeitsystemen werden Ausnahmen und ihre Behandlung jedoch häufig vernachlässigt. Auch Entwurfssprachen sollten über Möglichkeiten verfügen, Ausnahmen und Mittel zu ihrer Behandlung adäquat zu beschreiben. Insbesondere bei nicht-trivialen Exception-Handling-Mechanismen, wie sie etwa in verteilten Systemen notwendig werden können, ist eine formale Korrektheitsanalyse bzw. inhaltliche (simulationsgestützte) Validierung unerläßlich.

Petri-Netze werden immer häufiger als Beschreibungs- und Simulationssprache für Realzeitsysteme [CR85,DR85,HH89,Ige89,Pop89,Sch86] vorgeschlagen. Es existiert bereits eine Vielzahl von Werkzeugen, die den Einsatz von Petri-Netzen in der Praxis unterstützen, etwa zum graphischen Editieren, zur Simulation und zur Analyse [Fel90,LE89].

In dieser Arbeit werden auf Petri-Netzen basierende neuartige Konzepte zur Modellierung und Validierung von Exception-Handling-Mechanismen für Realzeitsysteme vorgestellt. Mechanismen zur Behebung von Ausnahmesituationen können in Petri-Netzen wie andere "reguläre" Systemabläufe beschrieben werden. Das Problem liegt in der Einbettung solcher Exception-Handling-Netze in den Gesamtsystementwurf, ohne diesen zu unübersichtlich werden zu lassen. Es ist sinnvoll, Exception-Handling-Netze getrennt von den Netzen zur Beschreibung des Normalverhaltens eines Systems zu untersuchen.

Trotzdem sollte es möglich sein, bei der Simulation von Systemabläufen auch Ausnahmesituationen und ihre Behandlung zu berücksichtigen und zu validieren. Eine entsprechende Erweiterung für einen Petri-Netz-Simulator [OSL90] wird vorgestellt.

Wir betrachten hier Ausnahmen nicht nur als unerwünschte, unvorhergesehene Fehlersituationen sondern auch als Konzept zur Strukturierung von Systemspezifikationen. Damit wird ermöglicht, zwischen regulärem Systemverhalten und speziellen Sonderfällen zu unterscheiden. Die Lesbarkeit von Systemspezifikationen kann so erheblich verbessert werden.

Diese Arbeit ist wie folgt gegliedert: Im folgenden zweiten Abschnitt wird die Verwendung von Petri-Netzen zur Beschreibung von Realzeitsystemen motiviert und erläutert. Der dritte Abschnitt führt deklarative Beschreibungskonzepte für Ausnahmesituationen ein, und der vierte Abschnitt geht auf die Validierungsproblematik ein. Den Abschluß bildet ein kurzer Überblick über bereits durchgeführte Implementationsarbeiten.

2 Petri-Netze zur Beschreibung von Realzeitsystemen

Petri-Netze sind ein graphisch-formales Beschreibungsmittel für Systeme mit aktiven Systemkomponenten (z.B. Prozessen) und passiven Systemkomponenten (z.B. Betriebsmittel für Prozesse). Aktive Systemkomponenten werden als Transitionen (graphisch als Vierecke) und passive Systemkomponenten als Stellen (graphisch als Kreise) dargestellt. Eine Vorgänger-/Nachfolger-Beziehung (graphisch als Pfeile zwischen Stellen und Transitionen dargestellt) beschreibt die als Signal-, Daten- oder sonstige Objektflüsse gegebenen kausalen Zusammenhänge.

Systemzustände werden durch Marken in den Stellen modelliert. Transitionen repräsentieren Zustandsübergänge: Wenn eine Transition *schaltet*, dann werden aus ihren Eingangs-Stellen Marken entnommen und in ihre Ausgangs-Stellen Marken abgelegt.

Eine Transition ist *aktiviert*, d.h. sie kann schalten, wenn in ihren Eingangs-Stellen genügend Marken (entsprechend den jeweiligen Kantenbeschriftungen) vorhanden sind, und die Kapazität ihrer Ausgangs-Stellen zur Aufnahme von Marken (entsprechend den jeweiligen Kantenbeschriftungen) ausreicht. Ein logischer Ausdruck als Inschrift der Transitionen ermöglicht (optional) die Formulierung zusätzlicher Bedingungen an die beim Schalten verwendeten bzw. erzeugten Marken.

Beispiel 1:

Abbildung 1 zeigt einen Ausschnitt aus einem höheren Petri-Netz (Prädikate/Transitionen-Netz) zu einem einfachen Steuerungsvorgang, bei dem ein Ventil geöffnet bzw. geschlossen werden kann. Öffnen und Schließen des Ventils werden als Transitionen modelliert, die den Status der Stelle "Ventilzustand" von "geschlossen" nach "offen" bzw. umgekehrt verändern. Der aktuelle Zustand des Ventils wird durch eine Marke in der Stelle "Ventilzustand" angezeigt. Zusätzliche Einflußgrößen auf die Transitionen "Ventil-öffnen" bzw. "Ventil-schließen" wie z.B. Druck- und Temperaturanzeige sind zur Vereinfachung weggelassen worden.

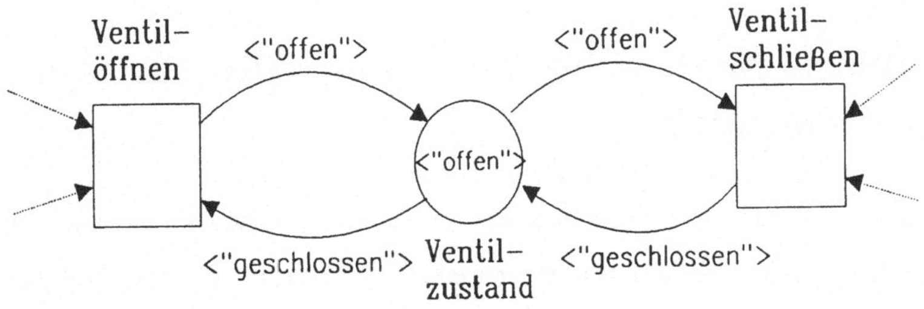

Abbildung 1: Einfaches Prädikate/Transitionen-Netz

Petri-Netze sind besonders geeignet zur Modellierung von verteilten Systemen. Formale Konzepte zur Vergröberung/Verfeinerung von Netzen und die damit mögliche Erstellung von Netz-Hierarchien erlauben die Betrachtung eines Systems auf verschiedenen Abstraktionsstufen, und unterstützen die schrittweise Beschreibung von umfangreichen Systemen in einer Top-Down- oder Bottom-Up-Vorgehensweise.

Die Verwendung unterschiedlicher Netz-Typen ermöglicht einen schrittweisen Übergang von eher informalen Systembeschreibungen in Form von Kanal/Instanzen-Netzen zu formalen Systembeschreibungen in Form von Prädikate/Transitionen-Netzen [Rei85]. Für letztere sind formale Analysen ebenso möglich wie simulationsgestützte Validierung. Grundlagen der Petri-Netz-Theorie und Einführungen in verschiedene Anwendungsbereiche vermitteln beispielsweise [Bau90,Rei85,Rei86].

In [CR85,DR85,HH89,Ige89,Pop89,Sch86] werden Petri-Netze als Spezifikationssprache für Realzeitsysteme verwendet. Es sind verschiedene Erweiterungen für Petri-Netze zur Modellierung von Zeitaspekten vorgeschlagen worden, Übersichten dazu geben [Eco87,Obe90c]. Wir verwenden einen Ansatz aus [DR85], der es ermöglicht, lokale

Uhren (wie sie für verteilte Systeme typisch sind, in denen jeder Rechner über eine eigene Uhr verfügt) und Synchronisationsmechanismen zwischen diesen in Netzen zu beschreiben. Es können so beispielsweise modelliert werden:

- Uhrzeiten,

- Zeitdauern, Zeitabstände,

- Fristen, Termine,

- relative zeitliche Zusammenhänge.

Eine ausführliche Beschreibung findet sich in [Obe90c].

3 Deklarative Beschreibungskonzepte für Ausnahmesituationen

Als Ausnahmen werden hier Systemzustände bezeichnet, in denen eine sog. *Ausnahmebedingung* erfüllt ist. Ausnahmen stellen einmal Abweichungen von im allgemeinen geltenden Regeln (*Normen*) dar. Ausnahmen können aber auch als Abstraktionsprinzip verwendet werden, um bei der Beschreibung des Normalverhaltens eines Systems selten auftretende Spezialfälle vernachlässigen zu können.

Ausnahmebedingungen können sich auf einzelne Zustände oder auf Zustandsübergänge beziehen. Sei beispielsweise folgende Regel gegeben: "Wenn die Temperatur 120°C übersteigt, dann muß das Ventil geöffnet sein". Die dazugehörige zustandsbezogene (*statische*) Ausnahmebedingung lautet: "Das Ventil ist geschlossen, und die Temperatur übersteigt 120°C". Eine weitere Regel sei: "Wenn die Temperatur 150°C überschritten hat, dann darf das Ventil erst geschlossen werden, wenn die Temperatur unter 100°C gesunken ist". Die entsprechende zustandsübergangsbezogene (*dynamische*) Ausnahmebedingung lautet: "Die Temperatur hat 150°C überschritten, und danach ist das Ventil geschlossen worden, bevor die Temperatur unter 100°C gesunken ist ".

Eine Besonderheit stellen zeitbezogene Ausnahmebedingungen dar, beispielsweise:

- Ein Prozeß P dauert weniger als δ_1 oder mehr als δ_2 Zeiteinheiten.

- Ein Betriebsmittel B wird zu einer bestimmten Zeit λ benötigt, zu der es nicht verfügbar ist.

- Ein Prozeß P findet vor (nach) einem Termin τ statt.

- Zwischen Prozeß P_1 und P_2 liegen weniger (mehr) als δ Zeiteinheiten.

Allgemein geltende Regeln für das Systemverhalten können in Petri-Netzen durch sogenannte *Fakt-Transitionen* und *ausgeschlossene Transitionen* deklarativ formuliert werden [Obe90c].

- **Fakt-Transitionen** [GL78]

  repräsentieren Anforderungen an Markierungen der Stellen in einem Netz: Nur solche Markierungen (Systemzustände) sind zulässig, in denen keine Fakt-Transition aktiviert ist. Mit Fakt-Transitionen können etwa Regeln der Art "In keiner Markierung dürfen die Stellen S_1 und S_2 gemeinsam markiert sein" oder "Wenn Stelle S_1 markiert ist, dann muß auch Stelle S_2 markiert sein" modelliert werden. Graphisch werden Fakt-Transitionen als Rechtecke mit eingeschriebenem stilisierten F dargestellt.

- **ausgeschlossene Transitionen** [Vos87]

  repräsentieren Anforderungen an Markierungsfolgen (Zustandsfolgen). Es sind solche Markierungsfolgen ausgeschlossen, die Markierungen M^1 und M^2 enthalten, für die gilt:

  - M^1 liegt in der Markierungsfolge vor M^2.

  - Unter M^1 ist die ausgeschlossene Transition aktiviert (entsprechend der üblichen Petri-Netz-Schaltregel).

  - In M^2 hat die ausgeschlossene Transition geschaltet, d.h. aus den Eingangs-Stellen sind Marken entnommen worden, und in die Ausgangs-Stellen sind Marken abgelegt werden.

Es können so beispielsweise Regeln der folgenden Art modelliert werden: "Wenn in einer Markierung eine Stelle S_1 markiert ist, dann darf in keiner Folgemarkierung S_2 markiert sein". Mehrere ausgeschlossene Transitionen können miteinander kombiniert werden, um komplexere Regeln zu modellieren [Obe89], etwa: "Wenn in einer Markierung M_1 die Stellen S_1 und S_2 markiert sind und in einer späteren Markierung M_2 die Stelle S_3, dann muß in jeder weiteren Folgemarkierung S_4 markiert sein". Graphisch werden ausgeschlossene Transitionen als Rechtecke mit eingeschriebenem X (für *excluded transition*) dargestellt.

In den genannten Beispielen können $S_1, \ldots, S_4$ auch zeitliche Bedingungen darstellen, etwa eine konkrete Uhrzeit betreffend.

Eine Ausnahmebedingung ist dann erfüllt, wenn eine durch eine Fakt-Transition oder ausgeschlossene Transition repräsentierte Regel verletzt wird, d.h. wenn eine Fakt-Transition aktiviert ist, oder wenn eine ausgeschlossene Transition geschaltet hat. Eine solche Regelverletzung wird durch die Markierung von speziellen Stellen signalisiert.

Für jede Ausnahmebedingung wird ein Mechanismus benötigt (möglicherweise reicht ein Default-Mechanismus, z.B. *Backward-Recovery* aus), der dann abläuft, wenn die betreffende Bedingung erfüllt wird. Zu jeder Fakt-Transition und jeder ausgeschlossenen Transition wird deshalb ein Netzfragment angegeben, das die Abläufe modelliert, die ausgelöst werden müssen, wenn die betreffende Fakt-Transition aktiviert ist oder die betreffende ausgeschlossene Transition stattgefunden hat.

Externe und interne Ausnahmebedingungen (Interrupts und Signale [HH89]) können so gleichartig behandelt werden.

Exception-Handling-Mechanismen können in Petri-Netzen, ausgehend von einem Petri-Netz-Grobentwurf, in dem das reguläre Systemverhalten beschrieben wird, in einer dreistufigen Vorgehensweise schrittweise modelliert werden:

Schritt 1:

Die Ausnahmebedingungen werden informal bzw. semiformal in einem Kanal/Instanzen-Netz beschrieben. Der dazugehörige Exception-Handling-Mechanismus wird als einzelne Transition (im Sinne einer "Black-Box") angegeben.

Schritt 2:

Die Ausnahmebedingungen aus Schritt 1 werden mit Fakt-Transitionen bzw. ausgeschlossenen Transitionen formal beschrieben. Es können bestimmte Parameter an die Exception-Handling-Transition übergeben werden, um so unterschiedliche klärende Mechanismen in Abhängigkeit von diesem Parameter vorsehen zu können.

Schritt 3:

Die Exception-Handling-Transition aus Schritt 1 und 2 wird durch ein Prädikate/Transitionen-Netz verfeinert, das die betreffende Ausnahmebehandlung vollständig und formal beschreibt.

Beispiel 2:

Die einfache zustandsbezogene Ausnahmebedingung "Das Ventil ist geschlossen und die Temperatur ist höher als 120°C" wird zunächst, wie in Abbildung 2 gezeigt, als Stelle in einem Kanal/Instanzen-Netz mit einer informalen Beschriftung modelliert. Die Transition "EH1" repräsentiert einen noch näher zu spezifizierenden Exception-Handling-Mechanismus. Falls die Stelle markiert ist, dann wird "EH1" ausgelöst.

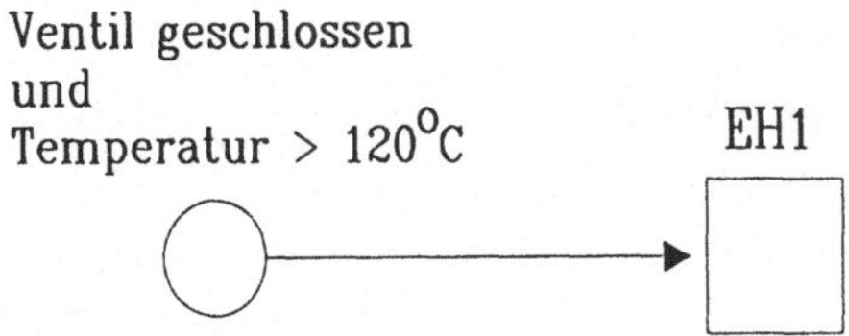

Abbildung 2: Zustandsbezogene Ausnahmebedingung

Zur formalen Beschreibung der zustandsbezogenen Ausnahmebedingung aus Abbildung 2 wird in Abbildung 3 eine Fakt-Transition "FT1" eingeführt, die als Eingangsstellen "Temperatur" und "Ventilzustand" besitzt. Falls FT1 schaltet, dann ist die Ausnahmebedingung erfüllt, und die Exception-Handling-Transition "EH1" ist aktiviert.

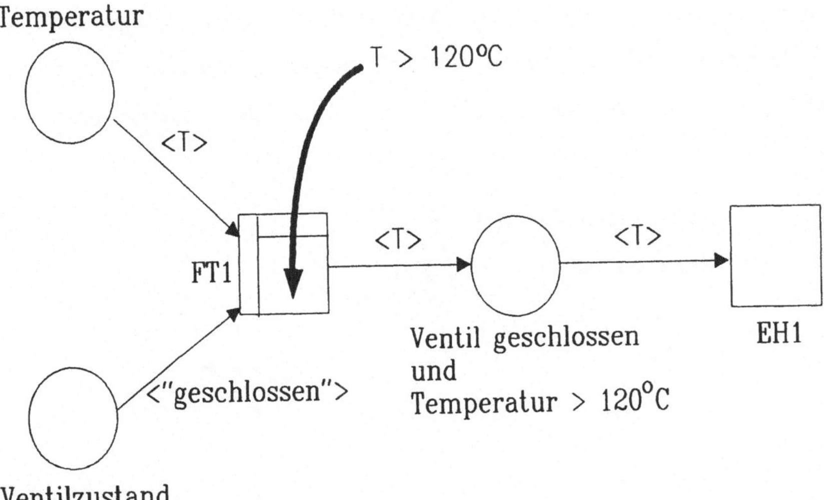

Abbildung 3: Fakt-Transition zur Ausnahmebedingung in Abbildung 2

Der aktuelle Temperaturwert wird an "EH1" als Parameter übergeben, d.h. "EH1" kann in Abhängigkeit vom Temperaturwert unterschiedlich aussehen. Bei Temperaturen über

140°C beispielsweise wird ein Sicherheitsventil geöffnet, bei Temperaturen über 150°C wird zusätzlich eine Kühlung eingeschaltet. Das entsprechende Netz zeigt Abbildung 4.

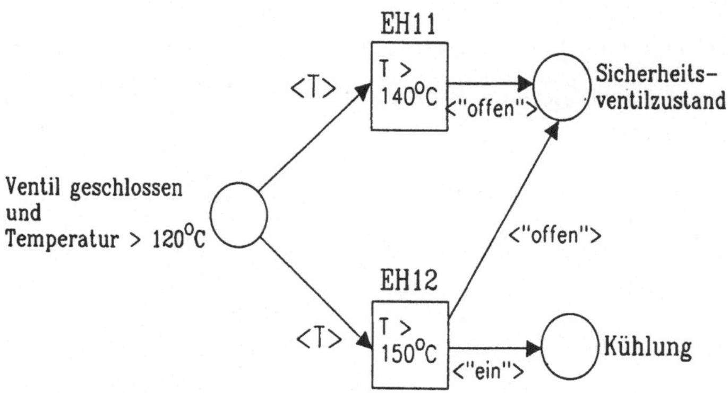

Abbildung 4: Verfeinerung der Exception-Handling-Transition aus Abbildung 2

Beispiel 3:

Abbildung 5 zeigt die informale Beschreibung der zustandsübergangsbezogenen Ausnahmebedingung "Die Temperatur hat 150°C überschritten, und danach ist das Ventil geschlossen worden bevor die Temperatur unter 100°C gesunken ist".

Wir verwenden hier eine standardisierte Schreibweise, nach der Anfangs- und Endzustand ("Alter Zustand" bzw. "Neuer Zustand") eines ausgeschlossenen Übergangs getrennt angegeben werden. Bei Bedarf können auch Zwischenzustände berücksichtigt werden.

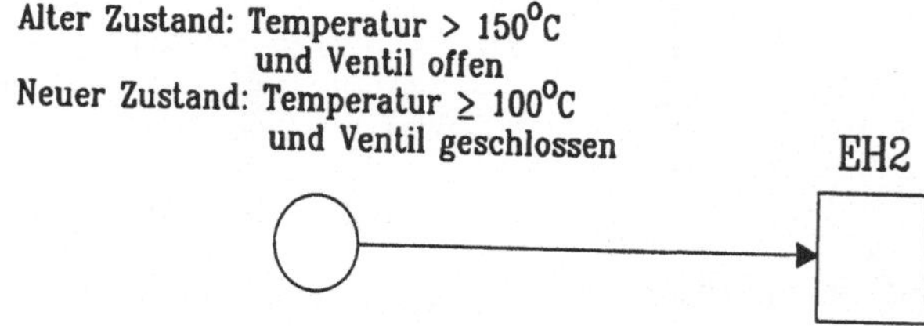

Abbildung 5: Zustandsübergangsbezogene Ausnahmebedingung

Für die zustandsübergangsbezogene Ausnahmebedingung aus Abbildung 5 wird eine ausgeschlossene Transition wie in Abbildung 6 gezeigt eingeführt.

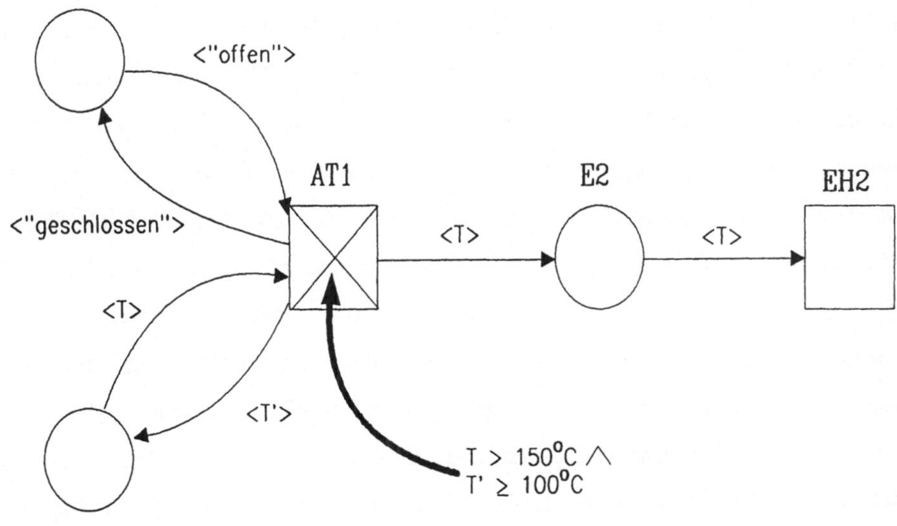

Abbildung 6: Ausgeschlossene Transition zur Ausnahmebedingung in Abbildung 5

□

Falls in der informalen Darstellung einer dynamischen Ausnahmebedingung Zwischenzustände enthalten sind, dann müssen mehrere ausgeschlossene Transitionen miteinander kombiniert werden [Obe90a].

Es können verschiedene Fortsetzungsmöglichkeiten für Ausnahmefälle modelliert werden (vgl. [Fed90]):

- Abbruch eines Prozesses,

- Fortsetzung eines Prozesses an der Fehlerstelle,

- Neustart eines Prozesses.

Ein Vorteil der vorgestellten Beschreibungstechnik für Ausnahmebedingungen besteht darin, daß für eine gegebene Ausnahmebedingung unmittelbar erkennbar ist,

- welche "passiven Objekte" (Maschinen, Personen, Material, ...) und

- welche "aktiven Objekte" (Prozesse, Operationen, Tätigkeiten, ...)

beteiligt sind.

In der Verfeinerung der Exception-Handling-Transition kann sofort festgestellt werden, welche passiven und aktiven Objekte vom Exception-Handling betroffen sind. Diese

Informationen sind beispielweise wichtig, wenn parallel ablaufende Prozesse auf die gleichen passiven Objekte zugreifen. Dann müssen in Verbindung mit dem Auftreten von Ausnahmen Sperrmechanismen eingesetzt werden, so daß kontrolliertes Exception-Handling möglich wird. Insbesondere der letztere Aspekt ist in der Literatur noch nicht ausreichend untersucht worden. Es sind hier ähnliche Konzepte denkbar, wie sie auch im Datenbankbereich zur Mehrbenutzerkontrolle verwendet werden (vgl. dazu [Obe88]).

Für umfangreiche Exception-Handling-Mechanismen kann es möglich sein, daß bei der Klärung einer Ausnahmesituation neue Fehlersituationen auftreten. Innerhalb eines Exception-Handling-Mechanismus können deshalb wiederum Ausnahmebedingungen definiert werden. Demgegenüber wird beispielsweise in [Mey88] angenommen, daß Mechanismen zur Ausnahmebehandlung leicht überprüfbar sind, und daher wird dort verlangt, daß diese Mechanismen in jedem Fall korrekt arbeiten. Diese Forderung ist jedoch für Realzeitsysteme im allgemeinen nicht realistisch.

4 Simulationsgestützte Validierung von Systementwürfen

Formale Analysen sind nur für relativ kleine Netze mit aktzeptablem Rechenzeitaufwand durchführbar. Für größere Netze bieten sich simulationsgestützte Validierungsmethoden an [OSL90].

Im Zusammenhang mit Exception-Handling interessiert insbesondere die Frage, ob alle relevanten Ausnahmesituationen beschrieben worden sind, und ob die im Entwurf vorgesehenen Mechanismen zu ihrer Behandlung ausreichend und korrekt sind. Es sind zwei Konzepte zur Simulation vorgesehen:

- Simulation des regulären Systemverhaltens unter der Vorgabe, daß keine Ausnahmesituation eintreten darf. So kann beispielsweise überprüft werden, welche Prozesse wann stattfinden müssen, so daß alle zeitlichen Restriktionen eingehalten werden und keine Ausnahmebedingung erfüllt wird.

- Simulation mit Einbeziehung der Mechanismen zur Behandlung von Ausnahmesituationen. Es können beliebige Ausnahmesituationen simuliert werden, um die vorgesehenen Exception-Handling-Mechanismen zu validieren.

Zur Überprüfung zustandsbezogener Ausnahmebedingungen reicht es aus, den jeweils aktuellen Systemzustand zu betrachten. Zustandsübergangsbezogene Ausnahmebedingungen können demgegenüber nur überwacht werden, wenn Informationen über vergangene Zustände vorliegen. Dazu können in zusätzlichen Stellen beim Eintreten bestimmter Ereignisse Marken als (parametrisierte) Statusanzeiger abgelegt werden.

In Beispiel 3 aus dem vorangehenden Abschnitt könnte etwa eine Stelle "Temperatur-Warnung" eingeführt werden, die dann markiert wird, wenn die Temperatur 150°C überschreitet. Diese Marke wird entfernt, wenn die Temperatur unter 100°C sinkt. Es kann die zustandsbezogene Ausnahmebedingung "Die Stelle "Temperatur-Warnung" ist markiert, und das Ventil ist geschlossen" formuliert werden, die äquivalent ist zur ursprünglichen dynamischen Ausnahmebedingung.

Simulation kann nach Beendigung von Schritt 2 bzw. Schritt 3 der in Abschnitt 3 angegebenen Vorgehensweise durchgeführt werden. Nach Schritt 2 ist Eingabe zur Simulation ein Petri-Netz zusammen mit Fakt-Transitionen und ausgeschlossene Transitionen. Falls eine Ausnahmebedingung erfüllt wird, dann wird diese durch Schalten einer Transition bereinigt.

Falls nach Schritt 3 zusätzlich eine Menge von Netz-Fragmenten vorliegt, die jeweils einen Exception-Handling-Mechanismus verfeinern, dann kann der genaue Ablauf des Exception-Handling simuliert werden.

Während eines Simulationslaufs ist es leicht möglich, Ausnahmebedingungen in Form von Fakt-Transitionen oder ausgeschlossenen Transitionen einzufügen, zu ändern oder zu löschen, d.h. die Schritte 2 und 3 können mehrfach durchlaufen werden.

5 Implementation

Das hier beschriebene Verfahren zur Modellierung und simulationsgestützten Validierung von Exception-Handling-Mechanismen mit Petri-Netzen wurde in ein existierendes Werkzeug zur Simulation und Analyse von höheren Petri-Netzen (Prädikate/Transitionen-Netze) [Obe88,Obe89,OSL90] integriert. Als Programmiersprache wurde Prolog [CM84] verwendet. Dabei konnte der Backtracking-Mechanismus von Prolog eingesetzt werden zum automatischen Zurücksetzen einer oder mehrerer Transitionen in Verbindung mit der Ausnahme-Behandlung.

Die beschriebenen Konzepte können aber auch in Verbindung mit anderen Netz-Simulatoren ohne Möglichkeit zum automatischen Zurücksetzen von Transitionen eingesetzt werden. Dann muß allerdings für jede Transition t eine kompensierende Transition $\hat{t}$ definiert werden, die die Wirkungen von t rückgängig macht (evtl. in Verbindung mit zusätzlich einzuführenden Stellen [Obe88]).

Die vorgestellten Konzepte sind auch Bestandteil des Methoden- und Tool-Pakets INCOME[1] [LNO*89,Sch89], das vor allem zur Entwicklung von Systemen in der Prozeß-automatisierung eingesetzt wird.

Danksagung

Für hilfreiche Diskussionen danke ich meinem Kollegen Tibor Németh.

Literatur

[Bau90]   B. Baumgarten. *Petri-Netze. Grundlagen und Anwendungen.* BI-Verlag, Mannheim, 1990.

[CM84]   W.F. Clocksin und C.S. Mellish. *Programming in Prolog.* Springer-Verlag, Berlin, Heidelberg, New York, 1984.

[CR85]   J.E. Coolahan und N. Roussopoulos. A Timed Petri Net Methodology for Specifying Real-Time System Timing. In *Proc. IEEE Int. Workshop on Timed Petri Nets, Torino/Italy*, S. 24–31, 1985.

[DR85]   M. Didic und G. Richter. Time & Clocks & Task Management. In *Proc. IEEE Int. Workshop on Timed Petri Nets*, S. 116–125, Torino/Italy, 1985.

[Eco87]   P. Economopoulos. Petri Nets: A Model for the Analysis of the Behaviour and Performance of Concurrent Systems. In F.H. Lochovsky, Hrsg., *Office and Data Base Systems Research '87*, S. 99–132, University of Toronto, July 1987.

[Fed90]   C. Feder. *Ausnahmebehandlung in objektorientierten Programmiersprachen.* Springer-Verlag, IFB 235, Berlin, Heidelberg, New York, 1990.

[Fel90]   F. Feldbrugge. Petri Net Tool Overview 1989. In G. Rozenberg, Hrsg., *Advances in Petri Nets 1989, LNCS424*, S. 151–178, Springer-Verlag, Berlin, Heidelberg, New York, 1990.

[GL78]   H.J. Genrich und K. Lautenbach. Facts in Place/Transition-Nets. In *Mathematical Foundations of Computer Science 1978*, Springer-Verlag, Berlin, Heidelberg, New York, 1978.

[HH89]   R.G. Herrtwich und G. Hommel. *Kooperation und Konkurrenz.* Springer-Verlag, Berlin, New York, Tokyo, 1989.

[Ige89]   B. Igel. Applikative Beschreibungsmethoden in verteilten Prozeß-steuerungen. In R. Henn und K. Stieger, Hrsg., *Proc. PEARL 89-Workshop über Realzeitsysteme*, S. 172–195, Springer-Verlag, Berlin, Heidelberg, New York, 1989.

---

[1]Produkt der PROMATIS Informatik GmbH & Co. KG, Straubenhardt

[LE89]   M. Leszak und H. Eggert. *Petri-Netz-Methoden und -Werkzeuge. IFB 197*, Springer-Verlag, Berlin, Heidelberg, New York, 1989.

[LNO*89]   G. Lausen, T. Németh, A. Oberweis, F. Schönthaler und W. Stucky. The IN-COME Approach for Conceptual Modelling and Prototyping of Information Systems. In *CASE89. The First Nordic Conference on Advanced Systems Engineering, Stockholm*, Mai 1989.

[Mey88]   B. Meyer. *Object Oriented Software Design*. IEEE Computer Society Press, New York, 1988.

[Obe88]   A. Oberweis. Checking Database Integrity Constraints while Simulating Information System Behaviour. In *Proceedings of 9th European Workshop on Application and Theory of Petri Nets (Venice/Italy)*, S. 299–308, Juni 1988.

[Obe89]   A. Oberweis. Integritätsbewahrendes Prototyping von verteilten Systemen. In M. Paul, Hrsg., *Proc. GI 19. Jahrestagung Computergestützter Arbeitsplatz, München, Bd. 1, IFB 222*, S. 215–230, Springer-Verlag, 1989.

[Obe90a]   A. Oberweis. Deklarative Systembeschreibungen als Grundlage für die Simulation mit Petri-Netzen. Proc. Symposium Simulationstechnik, Wien, 1990.

[Obe90b]   A. Oberweis. Die Behandlung von Ausnahmen in Software-Systemen: Eine Literaturübersicht. Forschungsbericht, Institut für Angewandte Informatik und Formale Beschreibungsverfahren, Universität Karlsruhe, Oktober 1990.

[Obe90c]   A. Oberweis. Zeitstrukturen für Informationssysteme. Dissertation, Universität Mannheim, Fakultät für Mathematik und Informatik, 1990.

[OSL90]   A. Oberweis, J. Seib und G. Lausen. PASIPP: Ein Hilfsmittel zur Analyse und Simulation von Prolog-beschrifteten Prädikate/Transitionen-Netzen. Zur Veröffentlichung eingereicht, 1990.

[PA90]   P.C. Philbrow und M.P. Atkinson. Events and Exception Handling in PS-Algol. *The Computer Journal*, 33(2):108–125, 1990.

[Pop89]   M. Popall. Simulation und Rapid Prototyping für Realzeitanwendungen auf Spezifikationsebene. In R. Henn und K. Stieger, Hrsg., *Proc. PEARL 89-Workshop über Realzeitsysteme*, S. 231–243, Springer-Verlag, Berlin, Heidelberg, New York, 1989.

[Rei85]   W. Reisig. *Systementwurf mit Netzen. Springer Compass*, Springer-Verlag, Berlin, Heidelberg, New York, 1985.

[Rei86]   W. Reisig. *Petrinetze. Eine Einführung. Studienreihe Informatik*, Springer-Verlag, Berlin, Heidelberg, New York, 1986.

[Sch86]   E. Schnieder. *Prozeßinformatik. Einführung mit Petrinetzen*. Vieweg-Verlag, Braunschweig, Wiesbaden, 1986.

58

[Sch89]   F. Schönthaler. *Rapid Prototyping zur Unterstützung des Konzeptuellen Entwurfs von Informationssystemen.* Dissertation, Universität Karlsruhe (TH), 1989.

[Sti89]   K. Stieger. PEARL90. Die Weiterentwicklung von PEARL. In R. Henn und K. Stieger, Hrsg., *Proc. PEARL89 - Workshop über Realzeitsysteme,* S. 99–137, Springer-Verlag, 1989.

[Vos87]   K. Voss. Nets in Data Bases. In W. Brauer, W. Reisig und G. Rozenberg, Hrsg., *Petri Nets: Applications and Relationships to Other Models of Concurrency,* S. 234–257, Springer-Verlag, Berlin, Heidelberg, New York, 1987.

Run-Time Prediction for Hard Real-Time Programs

Matjaž Colnarič

University of Maribor
Faculty of Technical Sciences
Smetanova 17, 62000 Maribor
Yugoslavia

colnaric@uni-mb.ac.mail.yu

Abstract

Timing predictability of the hard real-time systems can only be assured if every feature of them is deterministic. In this paper, various system design issues are stressed, from the system and processor architecture, to the language and exception handling. In the field of architecture, some design recommendations are given and a timing predictability study of microprocessors is carried out. Language features and exception handling are specially emphasised. For certain topics in these fields some solutions to avoid non-determinism are proposed.

1. Introduction

In recent years a shift in hard real-time systems research interest could be observed. Not the speed of execution, but the timeliness of actions is required. The definition of hard real-time systems is closely related with the deadlines of processes: if the cost function of missing a deadline of a process is a jump function to a very high value at a certain point, we deal with a hard real-time system; if it is continuously rising, we have a soft real-time environment. Missing deadlines in safety critical systems may result in a disaster and possibly in a loss of human lives.

Determining the execution time of a process is the basis for any deadline scheduling or schedulability verification. It is a complex task, where every feature, that can influence the process execution timing, has to be taken into account. Many researchers deal with this problem only partially.

Therefore, in this paper we should like to highlight some issues which did not receive sufficient attention, yet. We begin with a survey of some closely related work. In section 3, global requirements for various issues are established. The subsequent sections deal with these topics. Special emphasis is devoted to hard real-time language predictability issues and to exception handling.

2. Related Work

Recently, the predictability of hard real-time systems became a research topic receiving increased attention. The approach, complexity and levels of interest, however, differ from one author to another. In this chapter, a survey of some related work is given.

In [13], an approach is detailed aimed towards forecasting in early stages of hard real-time software development whether programs will meet their deadlines or not, avoiding costly analysis and prototyping methods. The use of a specification language within a CASE tool is proposed to build an early representation of the final software. The basic idea is to animate this representation to get an idea of the timing behaviour. As design progresses, the estimation can improve, since the model becomes more precise. To analyse timing in the model, special time description language constructs were developed to determine triggering, deadlines and duration of the processes. Two analysis methods are proposed, open and closed loop testing. These methods enable early discovery of unsatisfactory design, but the precision depends on the designer's skill to estimate execution times.

In [18] a study of real-time programs' execution time estimation is given, being part of the MARS project [12]. The proposed pre-run-time scheduler is only possible under the assumption of knowing estimates for maximum run-times of processes, which are based on the deterministic behaviour of the underlying hardware and operating system. Similar restrictions as in [21] and some new constructs to aid control of loops were introduced. In [19], an environment is described, where, after the analysis, a timing tree is built to determine the timing behaviour, with a possibility of asserting different time hypotheses to allow for experimentation with the system.

At the University of Texas at Austin the computer aided software engineering environment SARTOR is being developed. Parts of it are CACAP (Core Assembly Code Analysis Phase) and TAL (Timing Analysis Language).

CACAP is introduced in [1]. On the basis of generated assembly code, it determines the timing behaviour of real-time programs. First, straight-line assembly code blocks are located. Exact hardware system properties are stored in a data base, which is used by a hardware simulator. With this simulator straight-line assembly code execution times are determined. Then, a directed flow graph is created, in which execution times of straight-line assembly code parts are represented by nodes, and connections between them by edges. Well known algorithms on graphs are used to find loops and their execution times. Loops are subsequently replaced by these times thus reducing the graph. Also, the shortest and the longest path in the graph can easily be found, determining the shortest and the longest execution time between two nodes.

For the user interface there are two possibilities: a manual graphics interface using windows, menus, icons etc. and an interpreter for TAL, a special timing analysis language, described in [1] and [2]. The second possibility is much more sophisticated, enabling both automatic generation as well as manual improvement of the TAL programs. Analysis using TAL can produce results as accurate as detailed knowledge on the system's behaviour is incorporated into the TAL programs. The methods developed by this team can produce results very close to the actual worst-case response times. However, the method is very complex and difficult to implement, regarding the hardware description data-base and the TAL language and, thus, not very realistic for the practical use.

In [11] and [21] the language Real-Time Euclid is described. It has been designed for real-time applications and therefore fully supports timing-consistent multiprogramming. It uses only predictable language and system structures to be able to predict the timing behaviour of the system and the feasibility of scheduling.

Stoyenko's scheduling analysis method [21] consists of two parts. By the front end of the analyser necessary times for the execution of processes are computed. For this purpose it is recursively building trees of segments and resolving them on the way back. Statistics computed this way are used in the second part of the analyser, where a special technique called frame superimposition is employed. With this technique all possible combinations of sequences of frames (processes) to be processed are generated and the worst-case parameters are recorded. If all processes meet their deadlines the schedule is feasible.

In [6], basic reflections about hard real-time programming is given. We shall only focus on timing specification issues, that are also stressed in [9] and [7]. The main idea is to bound every language construct that threatens to be unpredictable: *events* are taken care of by the event supervisor introduced by the EXPECT and AWAIT keywords; a shared resource or a synchronisation primitive can only be blocked for a certain amount of time; loop iterations are bounded to a specified number; the use of GOTOs is restricted to skips. Applying these restrictions yields the possibility to estimate the run-times of procedures, blocks etc.

Deadline driven scheduling is proposed, with deadlines being explicitly stated and run-times derived using the above ideas. Each task's control block should contain information about deadline, execution time and residual time to complete the task. On these data, calculations can be carried out to determine the feasibility of the schedule, and predefined actions (possibly graceful degradation) can be taken when there is a danger of not meeting requested deadlines.

Kershaw [10] describes the VIPER microprocessor for safety-critical applications developed at the Royal Signals and Radar Establishment. Its design is expressed in the formal hardware description language ELLA and has been validated using formal algebraic techniques. Its features exclude any source of unpredictability: for example arithmetic overflow halts the processor, there are no exceptions or interrupts etc. In [4] the exception-free language NewSpeak with strictly bounded data types is described, enabling the estimation of upper bounds of the computing error, memory requirements and execution time.

A useful discussion about exceptions and some other predictability issues is given by the group proposing the specifications for PEARL 90 in [20].

3. Requirements

In order to be able to predict the response time of a process, it is inevitable to assure the predictability of every part of it. Once a process is scheduled to run on the processor, we must be able to determine what actions will be taken as well as their timing. Scheduling of processes is another problem and will not be dealt with in this paper. We are assuming that scheduling is deadline driven.

- *System architecture:* System design must not impose any non-deterministic delays. Dangerous features like virtual memory, DMA mechanisms, bus arbitration etc. must be carefully designed or not implemented.

- *Processor architecture:* The exact behaviour of the processor we are using must be known. Modern microprocessors are complex, their design and internal behaviour is proprietary, so it is usually difficult to determine it. They are designed on the basis of parallel operation of independent internal units, what makes the determination of action sequences and their timing behaviour more complicated. Further sources of unpredictability are asynchronous protocols, not adequately designed peripheral units (non-deterministic response time), cache memories that are burst-loaded, rejected prefetches, exceptions and interrupts that are internally serviced etc.

- *Language issues:* Common real-time languages do not assure run-time predictability of programs. To improve them, some restrictions have to be applied concerning dynamic features and bounding of loops and delays. Further, some language constructs have to be extended and new ones introduced.

- *Exception handling:* Exceptions present a serious problem for predicting run-times of processes. Attempts have been made to prevent them by exceptionless hardware designs, strict bounding of data items and exhaustive testing of software at compile-time. In the reality they should be avoided as much as possible by means of fault tolerance, graceful degradation and intelligent interfaces.

4. Architectural Issues

In non-time-critical systems the main objective in architecture design is to achieve high speed and the best possible system utilisation. To overcome the problem of slow peripherals, direct memory access techniques were invented. To extend physical memory, virtual memory is used. For communications, the Ethernet protocol is broadly implemented.

Early real-time systems were built on the same principles, "to be as quick and efficient as possible". In modern hard real-time systems, the speed is not the main feature any more. What is really important is, that the processes meet their deadlines. To assure that, the architectural features must be predictable. Hence, the following issues must be taken care of:

- Classically designed DMA is delaying the execution time of processes by stealing memory cycles. There have been propositions to overcome that problem (see [5]). The question is, using modern fast peripheral devices, if there is still the need for this kind of data transfer. Disks, for example, have transfer rates of up to 8 Mbytes per second. That means that after a data block is found, it is better to halt the processor and transfer the entire block, than to steal cycles over a longer period of time, resulting in the same delay. The processor will be halted for the time necessary to transmit the block, which is known in advance and can be taken into account at compile-time.

- Virtual memory is another source of unpredictability. To be able to predict a response time of a process, the amount of physical memory it requires has to be allocated. This should not impose a big problem at the current state of memory technology and pricing.

- In distributed systems, a deterministic type of data transfer has to be implemented. The broadly used CSMA/CD, for example, is not suitable, because of its random collision resolution mechanism.

- A bus architecture must be carefully selected. If an asynchronous data transfer protocol is chosen, then the peripheral devices must respond in bounded time. They shall have local intelligence to resolve recoverable problems or to report them to others (see Exception Handling). For bus arbitration, a deterministic protocol is necessary.

- It is important that a system architecture is kept simple. In simple, clear architectures it is more likely that the dangers of non-determinism will be located and prevented. Only such architectures can be tested exhaustively. In complex architectures, it is usually difficult to foresee all the traps that can make it unpredictable.

It is essential, that the processor, being the most important part of an architecture, is fully deterministic in its timing behaviour. Besides the already mentioned asynchronous protocol, there are some issues that need to be stressed:

- *On-chip cache memory and instruction prefetch:* Modern universal microprocessors often include on-chip cache memory to increase execution efficiency by providing quick store for instructions, operands and data (see [14,15,16]). Another benefit of using a cache is that the processor's external bus activity is largely reduced.

  In microprocessors, program execution is often organised in form of a two- or three-stage pipeline. During execution of an instruction, the bus controller prefetches the next instruction. First, the cache is checked for that data. If a cache hit occurs, data are transferred through the instruction path, enabling the execution without any delay. If not, an external memory read cycle is performed and data are supplied to the decoder and (at no loss of time) written into the cache. A cache can also be filled in burst mode, where a bigger part of it is updated by data that is expected to be needed in the next cycles.

  In data sheets, instruction execution times are stated for both "hit" and "not hit" cache examples. Exact program execution time may be calculated, although it seems to be quite complicated, because the exact processor behaviour is usually not known publically. Calculating with worst-case data yields the worst-case result, which will not be too pessimistic in average programs with not too many short loops. Burst filling presents a problem: the delay, although deterministic, is difficult to foresee. The system apparently loses determinism and the feature should therefore be avoided.

- *Operand misalignment:* Prefetches are usually performed in long words to utilise the bus well. If the address of a long word operand falls across the long word boundary, multiple external bus cycles have to be used to fetch it.

- *Concurrency:* Modern processors allow internal units like the sequencer, the pipeline and the bus controller to operate independently (see [14,17]). If, for example, an instruction ends with a write to the external memory, the sequencer can already start decoding the next instruction, whose code has been prefetched. The sequencer may also request a bus cycle while the bus controller is busy. This cycle is executed after the completion of the current one (cp. interesting examples in [15] and [16]). That causes a sequencer to wait, delaying microcode execution. This delay is deterministic, although again impossible to calculate because of lack of information.

An example of a totally predictable processor is the 32-bit VIPER [10]. It was developed by the Royal Signals and Radar Establishment for use in safety-critical applications. Its design was expressed in formal terms in the hardware description language ELLA and validated using formal algebraic techniques. To be able to prove and guarantee its behaviour, they had to keep it as clean and simple as possible. Its logic is implemented in hardware instead of being microprogrammed. VIPER has none of the above stated problems. Furthermore, on every exception it reacts by halting, including arithmetic overflow or underflow, when desired. It does not support interrupts and suggests to implement polling. To program it, special languages have been developed: the structured assembler VISTA and NewSpeak (see [4]).

5. Language Issues

The aim of this section is to present some restrictions and extensions to be implemented in an existent high-level language. For this purpose, PEARL is chosen according to analyses given in the literature [6]. It has timing constructs that already enable real-time programming and present a certain basis for the proposed upgrade. Close references to this section are [11,21,9].

Some necessary restrictions need to be implemented to make the run-time programs predictable:

- *Dynamic features,* including recursion, dynamic data structures, references, dynamic procedure calls, virtual addressing etc. shall be forbidden, because otherwise unsolvable problems concerning time and memory space requirements arise.

- *Delays* caused by the synchronisation or waiting for resources must be bounded. This also comprises the run-time scheduling (see [11,9]).

- *Loops* must be bounded by compile-time known worst-case numbers of iterations stated by the programmer and are reduced to the execution time of the bodies multiplied by these counts plus the loop control overhead. If they are exceeded during the execution, a predefined action is taken.

- *GOTOs* can produce uncontrollable loops and are therefore limited to jumps to the start of subsequent constructs, simplifying exiting from inner structures. In run-time calculations, they can be ignored since the delay can only be shortened.

To make the execution times of high-level language programs predictable, the following solutions are proposed:

- *Straight-line sequences* of statements can be reduced to delays by calculating the execution times of the resultant assembly language code;

- *IF and CASE statements* are reduced to the execution times of the longest alternatives plus a control overhead delay. Only begins and ends of alternatives are marked, no additional language constructs are needed.

- *Loops* require a new construct:

loop-statement ::= [FOR id] [FROM int - - exp7] [BY int - - exp7] [TO int - - exp7]

[WHILE bit1 - - exp7] **MAXLOOP int - - exp7 EXCEEDING statement-sequence**

REPEAT block-tail

With the MAXLOOP parameter an additional loop counter is initialised to be decremented upon each iteration. If zero is reached, the EXCEEDING statement sequence is executed (to avoid an exception).

- *Explicit stating of execution time:* Only the worst-case execution times of all components can be specified, which are then summed and possibly multiplied, although they may be mutually exclusive. To avoid too pessimistic estimations, tighter execution times of certain parts of the task can be explicitly asserted if they are established and confirmed during the testing or through software re-use. Automatic analysis is then overridden. This estimation must be very reliable and some reserve is recommended. If it is exceeded during the execution, an action is taken.

To assert the known run-time, a new construct is proposed:

TIMEOUT-IN dur - - exp7 EXCEEDING statement-sequence

It should appear at the begin of any type of block. At compile-time, the automatic analysis is overridden and the given time is taken. An assembly code function is inserted, using the time-out stack, where a clock expression is placed. Entries in the stack are periodically decremented by a system timer service routine. If it runs out before the execution of the block is finished, the system behaviour is determined by the EXCEEDING statement sequence.

Recursive use of these rules yields the worst-case estimation of the run-time required by a task. For their implementation, two steps are required, a pre- and a post-processor.

In the *pre-processor*, the source code is analysed and typical constructs are marked by specially coded comments, which are then carried through the compiler to appear in the assembly source code, positioned at functionally the same places. Special meta-language constructs are considered and finally extracted from the program to be compiled on a standard PEARL compiler.

Assembly code generated by a PEARL compiler with included coded comments is input to the *post-processor*. It is parsed for the keywords and required actions are taken. Starting from the beginning, execution times of straight-line code for a certain processor (e.g. MC680XX) is computed and necessary functions are included. Execution times of system functions should be known and accessible from a library. Every loop, IF or CASE statement is reduced according to the above rules. Nested structures are reduced recursively.

6. Exception Handling

Exceptions are a common feature that seriously influences the behaviour of a system. As an exception, any request, that cannot be resolved within the context, where it was detected, is denoted. Events that trigger exceptions can be system and user interrupts, and hardware and software conditions that require a special system attention, like errors, warnings etc.

Interrupts are the means for interfacing the system with the real-world environment and per definitionem delay the execution of other processes. The alternative to them is polling, observing the state of some conditions and reacting on its changes. For example, the VIPER processor (see [10]) does not support interrupts to be more robust, testable and predictable. Systems based on common microprocessors can also employ polling. An example of a state observation based control system is described in [3]. If interrupts are supported, they have to be taken special care of. To retain predictability, a good idea is to migrate the administration of events to a parallel co-processor (see [8]). This paper will not deal with that problem.

The study of some compilers' run-time messages yielded the following classification of the possible sources of exceptions and their handling. A similar classification is given in [20]. We can divide the sources of exceptions into the following classes:

- exceptions caused by I/O operations,

- exceptions caused by invalid data,

- exceptions concerning independent parallel tasking,

- exceptions avoidable by appropriate restrictions, and

- the processor's system exceptions

a) Exceptions caused by I/O operations

- *Invalid device addressing (invalid unit identification, no such device):* In real-time programs, I/O units should be declared (for example as in the system part in PEARL) using compile-time known terms to allow for validation at compile-time. During the initialisation, these units are to be tested and eventual losses or failures reported. A failure during the operation, however, still remains a problem to be solved by the methods of fault tolerance (see below).

- *I/O device errors:* Intelligent, fault tolerant, self checking peripheral devices can recognise their malfunction. They shall have as much intelligence as possible to reasonably react on conflict situations. In the case of recoverable errors, they should try to recover locally. In the case standby redundancy is available, it can be switched over to the redundant units. In other cases they should report errors to the system, what inevitably disturbs the timing.

- *Exceptions caused by disk management:* These are special cases of I/O errors and are due to unknown, improperly addressed, locked, wrongly formatted or organised files, using wrong parameters, errors in I/O operations, problems with direct or keyed access, and reading more data than exist.

  In hard real-time systems, mass storage units are mainly needed for archiving and buffering purposes and, possibly, to provide knowledge bases. For archiving and buffering the sequential organisation is sufficient, and the use of a knowledge base in hard real-time systems is questionable; if it is implemented it still can be sequentially organised, although its use then turns to be rather awkward.

  Mass storage handling operations should therefore be limited, to eliminate as many problems as possible. Only sequential organisation is recommendable and only file names known at compile-time may be used. When reading a record, check for the existence of the next one should be made and the EOF flag should be returned when reading the last existent record. In the case of an error, a flag is returned, data shall have the value "undefined" (see below), and the user shall cope with the consequences.

b) Exceptions caused by invalid data:

- *Traps and errors concerning range overflow or underflow in arithmetic operations:* It is essential that the result of an operation is reasonable (i.e. as expected). That means that it is correct (if feasible) or the user is notified that it is not. One solution to this problem would be as is suggested in NewSpeak ([4]): to bound the data type declarations very strongly and allow only operations which do not leave the specified ranges. The problem is solved at compile-time, although it seems to be very restrictive.

  Our suggestion is to use the longest possible word to accommodate as large values as possible. If a result of an operation exceeds this limit, it should be taken as "undefined", represented by an asserted reserved value for every data type (for example, -0 for numerical values). The slight difference between "undefined" and "infinite" is that any operation on the "undefined" operand always yields an "undefined" result and does not create the false hope to be able to calculate with the "infinite" value. This means that somewhere in a calculation chain a non-solvable problem arose and the result is useless.

  I/O interfaces should have a certain degree of intelligence, so that if the final result that is output to them is "undefined", they react in a predefined manner. Maybe they can perform "graceful degradation" locally (if the action to be taken is not vital or can be pre-programmed for that case) or, if inevitable, report alert. Of course, the variables can be checked for the "undefined" value earlier in the computation and actions can be taken as soon as possible.

- *Arithmetic functions with illegal run-time parameter values (square root, logarithms etc.:* If it is possible (like in NewSpeak) to strictly declare bounds of parameters, they can be checked for potential violations. Since this seems often impractical, again the use of the "undefined" value for the same reasons as above is proposed.

- *Format declaration/run-time value mismatch or conversion error in I/O operations:* In this case the best possible feasible conversion is to be made. If a reasonable conversion is impossible, we have again an example to be solved by the "undefined" value (for example, entering non-numericals values when numerical ones are expected).

c) Exceptions concerning independent parallel tasking:

These problems arise in the case of conflict situations in tasking; for example, when a task intends to terminate, suspend or resume etc. a non-existent other task, or activate an already existent one etc. In the operating system we can pre-program reactions on some of these situations, but, according to the nature of parallel systems, all of them cannot be avoided (or it is dangerous to ignore them); however, they can be prevented by the programmer checking the current situation in the system.

d) Exceptions avoidable by appropriate restrictions:

In the above stated references and in this paper, several restrictions concerning features on each level of hard real-time systems' design are proposed. Observing them can eliminate a large number of run-time exceptions, like for example problems concerning dynamic or virtual addressing etc.

e) Processor's system exceptions:

Processors react on a number of hardware and software signals. For example, there is a bus error signal, indicating problems with data transfer and raising an exception when set; or, division by zero triggers a system exception. These exceptions can only be avoided by preventing their causes, like using a fault tolerant data transfer protocol or preventing a divisor to assume the value zero. However, all system exceptions cannot be avoided.

7. Conclusion

In the paper, predictability issues in hard real-time programs were stressed. Special emphasis was placed on language issues and exceptions. For these topics, some suggestions to enable or improve predictability were introduced.

In future work, architectural issues will be considered. As shown in chapter 4, the design of common microprocessors does not fulfill the requirements completely. To be able to make fully predictable hard real-time systems, a specially designed hardware, based on asymmetrical multiprocessor systems, will be proposed.

As the final result of the project, a totally predictable hard real-time system is expected.

References

[1] Rukmin Prasanna Amerasinghe. *A Comprehensive Compiler Based Timing Analysis Tool for Real-Time Software.* Master's thesis, University of Texas at Austin, May 1989.

[2] Moyer Chen. *The Dataflow Real-Time System: A Case Study.* Master's thesis, University of Texas at Austin, December 1987.

[3] Matjaž Colnarič, Ivan Rozman, Maksimilijan Gerkeš, and Bruno Stiglic. State-observation based hard real-time system. In *Proceedings of Software Engineering for Real-Time Systems*, IEE, Cirencester, September 1989.

[4] I. F. Currie. NewSpeak: a reliable programming language. In *High-integrity Software*, pages 122–158, Pitman Publishing, London, 1989.

[5] Wolfgang A. Halang. On methods for direct memory access without cycle stealing. *Micropro-cessing and Microprogramming*, 17:277–283, 1986.

[6] Wolfgang A. Halang. On real-time features available in high-level languages and yet to be implemented. *Microprocessing And Microprogramming*, 12:79–87, 1983.

[7] Wolfgang A. Halang and Richard Henn. On-Time: a high level language environment based on a portable operating system featuring feasible and fault tolerant scheduling. In *Proceedings of Software Engineering for Real-Time Systems*, pages 121–125, IEE, Cirencester, September 1989.

[8] Wolfgang A. Halang. Parallel administration of events in real-time systems. *Microprocessing and Microprogramming*, 24:687–692, 1988.

[9] Wolfgang A. Halang. A priori execution time analysis for parallel processes. In *Proceedings of the Euromicro Workshop on Real-Time*, pages 62–65, Como, June 1989.

[10] J. Kershaw. *The VIPER Microprocessor*. Technical Report 87014, Royal Signals and Radar Establishment, Malvern, Worcs. London: Her Majesties' Stationary, November 1987.

[11] Eugene Kligerman and Alexander Stoyenko. Real-time Euclid: a language for reliable real-time systems. *IEEE Transactions on Software Engineering*, 12(9):941–949, September 1986.

[12] Hermann Kopetz, A. Damm, Ch. Koza, M. Mulazzani, W. Schwabl, Ch. Senft, and R. Zain-linger. Distributed fault-tolerant real-time systems: the MARS approach. *IEEE Micro*, 9(1):25–40, February 1989.

[13] Rudolf J. Lauber. Forecasting real-time behaviour during software design using a CASE environment. *The Journal of Real-Time Systems*, 1(1):61–76, 1989.

[14] Doug MacGregor and Jon Rubinstein. A performance analysis of MC68020-based systems. *IEEE Micro*, December 1985.

[15] Motorola. *MC68020 32-bit Microprocessor User's Manual*. Prentice Hall, Inc., Englewood Cliffs, N.J., second edition, 1985.

[16] Motorola. *MC68030 Enhanced 32-bit Microprocessor User's Manual*. 1987.

[17] David Olsen. Effects of pipelining on algorithms for the MC68020. *Digital Design*, June 1985.

[18] Peter Puschner and Christian Koza. Calculating the maximum execution time of real-time programs. *The Journal of Real-Time Systems*, 1(2):159–176, 1989.

[19] Peter Puschner and Ralph Zainlinger. Developing software with predictable timing behaviour. *IEEE Real-Time Systems Newsletter*, 6(2):70–76, 1990.

[20] Klaus Stieger. PEARL 90 — Die Weiterentwicklung von PEARL. In R. Henn and K. Stieger, editors, *PEARL 89 — Workshop über Realzeitsysteme, 10. Fachtagung des PEARL-Vereins e.V. Informatik-Fachberichte 231*, pages 99 – 137, Springer-Verlag, Berlin-Heidelberg-New York, 1989.

[21] Alexander Stoyenko. *A Real-Time Language With A Schedulability Analyzer*. PhD thesis, University of Toronto, December 1987.

Ein portabler Zellencontroller

Rolf Blumenthal, Jürgen Geidies
Werum Datenverarbeitungssysteme GmbH, Lüneburg

1.   Einleitung

Im Rahmen der Entwicklung von CIM-Strukturen in der Fertigung ist der Zellencontroller das Bindeglied zwischen Maschinensteuerungen und Leitstand. Der Zellencontroller erfüllt die Forderungen der Betriebe nach DV-Unterstützung direkt an der Fertigungszelle und die der Planer nach normierten Kommunikationswegen von den Maschinen bis hin zu einem Leitstand.

Der vorliegende Aufsatz stellt – ausgehend von einem Einsatz des Zellencontrollers in einem Anwendungsfall der Automobilindustrie – wesentliche Funktionen und Eigenschaften des Systems vor. Dabei beschränkt er sich auf diejenigen Funktionen und Eigenschaften, die den Zellencontroller zu einem flexiblen Werkzeug für die Automatisierung vieler Anwendungen werden ließen.

2.   Ziele

Zur Entwicklung und Realisierung von CIM-Strukturen in der Fertigung wurde ein Zellencontroller konzipiert und realisiert. Der Zellencontroller ordnet sich in hierarchischen Automatisierungsstrukturen ein zwischen den Maschinensteuerungen und einem Fertigungsleitstand.

Es ist zwingend notwendig, an diesen fertigungsnahen Stellen Automatisierungstechnik einzusetzen, da die Fertigungsanlagen immer komplexer und größer werden und dadurch verstärkt den Charakter von autonomen Fertigungszellen besitzen. Dieser Charakter erzwingt Informationen über die Maschine und das Produkt direkt vor Ort. Diese Informationen können nicht von den Maschinensteuerungen selbst errechnet und dargestellt werden, wenn die zu automatisierende Zelle zu groß und unter Umständen mit sehr vielen Steuerungen bestückt ist. Sie können auch nicht insgesamt von einem Fertigungsleitstand erbracht werden, da dieser als übergeordnetes System für die gesamte Fertigung nicht alle notwendigen Details einer Zelle bearbeiten kann. Zusätzlich fördert der Zellencontroller auch die Autonomie der Zelle für die Zeiten, in denen der Leitstand nicht zur Verfügung steht.

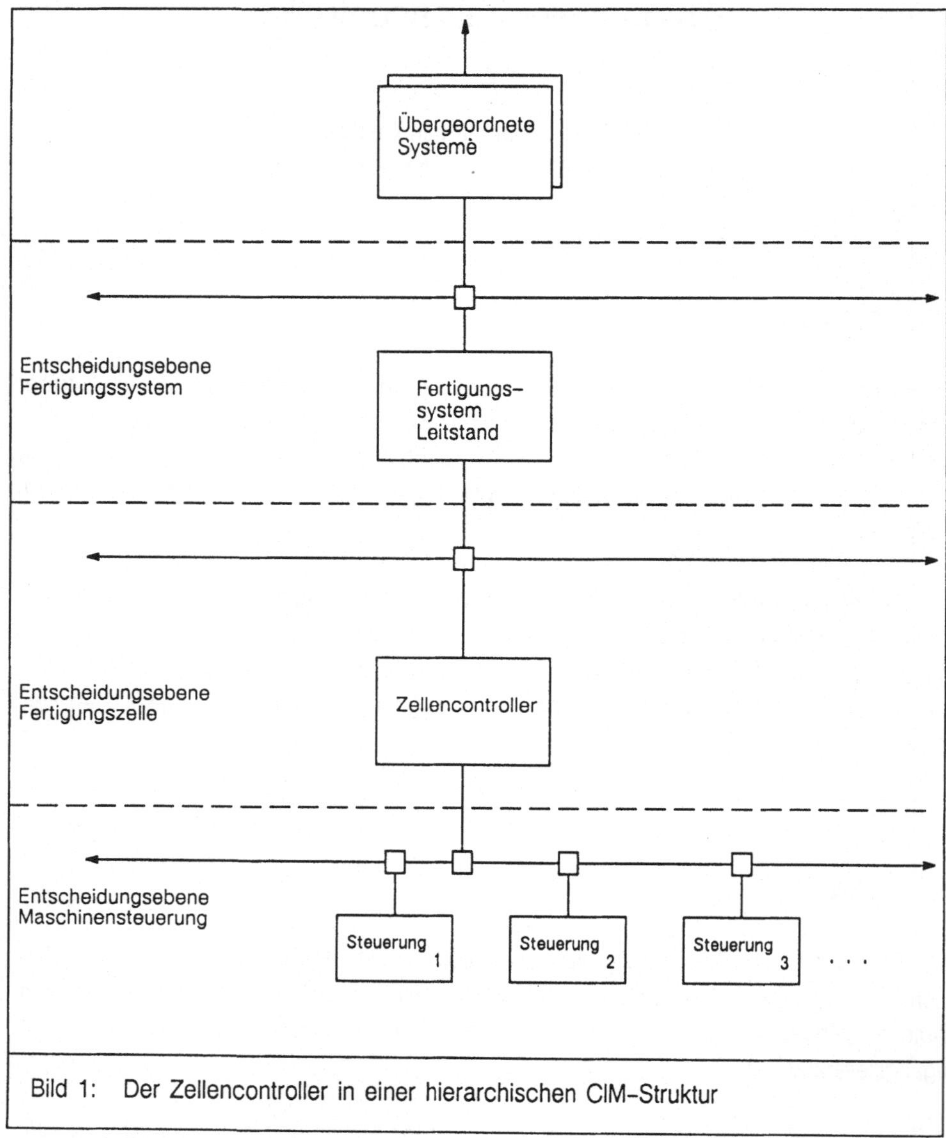

Bild 1:   Der Zellencontroller in einer hierarchischen CIM-Struktur

3.   Der Anwendungsfall

Das Konzept des Zellencontrollers und sein Funktionsumfang wurden zusammen mit Automobilfirmen erarbeitet, wobei die speziellen Anforderungen aus der Automobilfertigung wie z.B. taktorientierte Produktion besonders berücksichtigt wurden.

Als typischer Anwendungsfall wurde zuerst eine Fertigungszelle zur Herstellung von Sitzrahmen mit einem Zellencontroller ausgestattet. Dieser Zelle werden Rohre und Blechteile zugeführt. Durch Biege- und Schweißvorgänge entsteht das Gerüst eines Autositzes. Die Zelle selbst besteht aus einer Reihe von Schweißdrehtischen und Biegevorrichtungen.

Zwischen diesen Bearbeitungsaktionen sorgen Transport- und Handlingsysteme für den Fluß des Materials durch die Zelle. Die gesamte Ausdehnung der Zelle beträgt ca. 30 m x 150 m. Sie ist komplett umrandet mit Schutzgittern, die an vielen Stellen zur Beseitigung von Störungen zu öffnen sind.

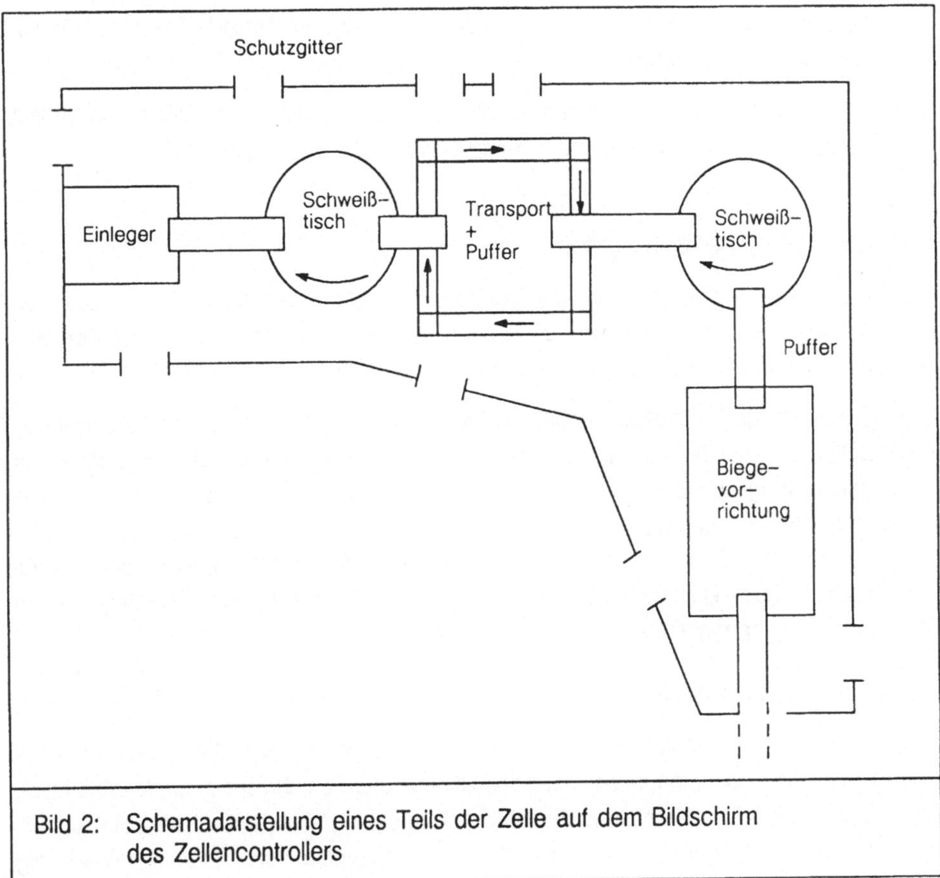

Bild 2:   Schemadarstellung eines Teils der Zelle auf dem Bildschirm
          des Zellencontrollers

Beim Betrieb der Fertigungszelle kann der Zellencontroller folgende Aufgaben behandeln:

o   Anlagenvisualisierung und Produktionsverfolgung

         In grafischen Darstellungen der Zelle können alle Ereignisse in der Zelle, der Zustand der Anlagen und der Materialfluß dargestellt werden.

         Im einzelnen liefert die Zelle Betriebszustände, Not-Aus, Schutzkreise, Anlagenstörung, Taktzeitüberschreitung, Materialmangel und -stau sowie Taktzeiten, Stückzahlen und Pufferfüllstände o   Unterstützung der Instandhaltung vor Ort

Alle Störungen innerhalb der Zelle sind jederzeit aktuell in Listenform abrufbar. Zur Unterstützung bei der Instandhaltung bei schwerwiegenden Störungen stehen ein Maßnahmenkatalog und eine Ersatzteil- und Werkzeugliste als Empfehlung für das Personal zur Verfügung. Um eine detaillierte Analyse der Störungen nach Fehlerursachen vornehmen zu können, müssen per Dialog vordefinierte Fehlerbegründungen eingegeben werden.

Der Einsatz dieser Funktionen empfiehlt sich insbesondere dann, wenn das Bedienpersonal und die Instandsetzer direkt bei der Zelle positioniert sind.

o   Überprüfung der Sollvorgaben

Da getaktete Anlagen über die Taktzeit eine relativ genaue Prognose zulassen, wieviele Teile produziert werden können, ist hier durch den Zellencontroller auch eine genaue Überprüfung der Sollvorgaben möglich.

Durch stetige Taktzeitüberwachung, Stückzählung und Störerkennung ist zu jedem Zeitpunkt einer Schicht die Gegenüberstellung von Produktionsverlauf und Sollvorgaben möglich. Sollten Sollvorgaben nicht eingehalten werden, ist dies exakt erklärbar und nachvollziehbar aufgrund der dokumentierten "Lebensgeschichte" der Anlagen. Dabei werden sowohl schwerwiegende, große Störungen als auch die Summe vieler kleiner, unbeachteter Verzögerungen der Nennausbringung der Anlagen entgegengestellt.

o   Dokumentation der Produktion

Anlagenführer, Schichtführer und Produktionsleiter erhalten über eine Reihe von Protokollen Dokumente über den Produktions- und Schichtverlauf. Dabei sind häufig sowohl detaillierte Rohinformationen zur Betrachtung eines kritischen Zeitabschnitts als auch Summenanalysen für die Trenderkennung und Planung notwendig. Aufgrund aller zur Verfügung stehenden Informationen kann der Zellencontroller von allen Bearbeitungsstationen und Anlagen für die Planung nützliche Kennwerte wie Nutzungsgrad, Verfügbarkeit oder mittlere Taktzeit errechnen. Die Betreiber der Zelle erhalten so fundierte Stationsbewertungen.

Entscheidende Aussagen für die Produktion sind natürlich die produzierten Stückzahlen, ggf. nach den unterschiedlichen Sorten gegliedert und dem Sollwert gegenübergestellt.

4. Funktionen des Zellencontrollers

Um den geschilderten Anforderungen gerecht zu werden, ist der Zellencontroller mit einer Reihe von Funktionen ausgestattet worden. Einige dieser Funktionen sollen hier vorgestellt werden:

Erfassung der Daten durch die Steuerungen

In größeren Fertigungsbereichen kann der Einsatz unterschiedlicher Steuerungen und damit verbunden unterschiedlicher Kommunikationsvarianten, erforderlich sein. Dies führte zu der Notwendigkeit, einen Zellencontroller parallel mit verschiedenen Steuerungen zu betreiben. Da Art und Anzahl dieser Variationen nicht festzulegen sind, bietet der Zellencontroller die Möglichkeit, die angeschlossenen Steuerungen per Parametrierung zu definieren.

Zur Kommunikation benutzt der Zellencontroller jeweils eine einheitliche Protokollschicht. Diese Schicht wird dann pro angeschlossener Steuerung auf die konkrete Kommunikationsschicht abgebildet.

Datenmodell

Mit Hilfe der einheitlichen, vom Zellencontroller definierten Protokollschicht werden Daten zwischen Steuerung und Zellencontroller ausgetauscht. Diese Daten können unterschiedliche Darstellung und Bedeutung besitzen. Um sie allgemein nach bestimmten Kriterien bearbeiten zu können, müssen diese Daten in Gruppen zusammengefaßt werden. Die Daten einer Gruppe gehören dann alle zu einem bestimmten Datentyp, der nach definierten Kriterien übertragen und bearbeitet wird. Es werden im einzelnen unterschieden:

o   Meldungen und Alarme
        zur Übermittlung von Betriebszuständen oder Störungen

o   Meßwerte
        zur Bearbeitung von Taktzeiten, Prozeßgrößen und Prüfmerkmalen

o   Zählwerte
        zur Stückzählung und Überwachung von Pufferfüllständen

o   Fehlersignale
        bei Verwendung von Fehlercodes in komplexen Überwachungssystemen

o   Datenstrings
        bei der Erfassung von Stückidentifikationen (Barcodes) oder Auftragsdaten

All diese Daten können auf unterschiedliche Weise erfaßt, verarbeitet und in Grafikbildern visualisiert werden. Auch die Produktionsaussagen ergeben sich zum großen Teil aus der typspezifischen Betrachtung der Daten.

Adressierung der Daten

Um bei allen Produktionsauswertungen und -analysen stets eine genaue Ortung und Zuordnung zu den realen Zellenbereichen durchführen zu können, werden alle angeschlossenen Signale hierarchisch geordnet und adressiert. Dadurch ist es möglich, Auswertungen auf einzelne Bearbeitungsstationen, auf größere Anlagen oder auf die gesamte Zelle zu beziehen. Teilweise sind Kennwerte oder Analysen auch nur auf bestimmter Betrachtungsebene sinnvoll.

So ist für die Instandhaltung die Baugruppe als kleinste austauschbare Einheit eine wichtige Information. Für die Bewertung einer Bearbeitungsstation ist jedoch die Station als kleinste operative Einheit das wichtigste Element. Störanalysen und Produktionszahlen sind häufig auf globaler Anlagen- oder Zellenebene angesiedelt.

Gibt ein Zellencontroller seine Informationen an einen übergeordneten Leitstand weiter, erhält die Identifikation auf Zellenebene die Bedeutung einer globalen Identifikation gegenüber allen weiteren Zellencontrollern des Leitstands.

Auswertungen

Der Zellencontroller enthält eine große Anzahl von Produktionsdatenauswertungen in Form von Grafiken und Protokollen. Als Protokoll ist verfügbar

o   eine Störhistorie, die in chronologischer Reihenfolge den Verlauf der Fertigung dokumentiert. Alle angefallenen Störungen und die gefertigten Stückzahlen im Stundentakt bilden die Information dieses Protokolls

o   ein Schichtprotokoll mit der Bewertung der Anlagen und Operationen. Als Information stehen Kennwerte wie Nutzungsgrad, Verfügbarkeit und mittlere Taktzeit sowie summierte Stördauern und -häufigkeiten der einzelnen Störmeldungen pro Bearbeitungsstation zur Verfügung. Zusätzlich sind Produktionszeiten, Pausenzeiten und gefertigte Stückzahlen dokumentiert.

<table>
<tr><td>Anlage</td><td>Anlagenbezogenes Schichtprotokoll</td><td>15.05.1989, 14:15 Uhr</td><td>Seite: 1</td></tr>
</table>

Ausgabezeitraum vom: 15.05.1989 05:30 Uhr   bis: 15.05.1989 14:00 Uhr   Schicht: 3

Rechnerfehlzeiten: . . . . . .

Nutzungsgrad Anlage: . . . . %

Verfügbarkeit Anlage: . . . . %

Gefertigte Stücke: Sorte 1: . . . . . min
                    Sorte 2: . . . . . min
                    Sorte 3: . . . . . min
                    Sorte 4: . . . . . min

Pausenzeiten: Pause 1: . . : . .   bis . . : . .   Stückzahl in Pause: . . . . . Stck
              Pause 2: . . : . .   bis . . : . .   Stückzahl in Pause: . . . . . Stck
              Pause 3: . . : . .   bis . . : . .   Stückzahl in Pause: . . . . . Stck
              Pause 4: . . : . .   bis . . : . .   Stückzahl in Pause: . . . . . Stck
              Pause 5: . . : . .   bis . . : . .   Stückzahl in Pause: . . . . . Stck

Störzeiten: Mit Handeingriff   : . . . . . min
           Ohne Handeingriff : . . . . . min
           Summe   : . . . . . min

Stillstandszeiten: Materialmangel   : . . . . . min
                 Weitere Ursachen : . . . . . min
                 Summe   : . . . . . min

Stationsbezogene Störaussagen

Arbeitsfolge Kurztext	Arbeitsfolgenbezeichnung Klartext	Anzahl Störungen	Summe Störzeiten	Nutzungsgrad %	Verfügbarkeit %	mittlere Taktzeit

Bild 3:   Schichtprotokoll-Layout

Übersichten und Trendaussagen sind in Form von grafischen Auswertungen verfügbar. Einen großen Raum nimmt wiederum die Analyse von Störungen in Form von Summen und Klassifizierungen ein. Aber auch die Produktionsergebnisse im Verlauf der Produktionszeit in der direkten Gegenüberstellung zu Sollvorgaben als Kurve darzustellen ist sehr aussagekräftig. Dabei können globale, vergleichende Betrachtungen aller Fertigungsanlagen einer Zelle zur Auffindung von Engpässen führen.

Bei allen Auswertungen führt die Ausnutzung der hierarchischen Anlagenadressierung dazu, mit denselben Algorithmen sowohl detaillierte Aussagen zu einem spezifischen Teil – das Schutzgitter 2 des Schweißtisches A wird am häufigsten geöffnet – als auch globale Bewertungen – die Ausfallzeit von Schweißtisch B ist größer als die von Schweißtisch A – zu gewinnen.

Darüberhinaus gewinnt der Zellencontroller eine noch größere Flexibilität dadurch, daß er dem Anwender auch überläßt, welche Signale zu einer Auswertung herangezogen werden. Beispielsweise erscheinen Stückzahlen im Schichtprotokoll nur dann, wenn der Anwender bei einem Zählwert parametriert hat: "Dieser Zählwert ist ein Stückzähler für Schweißtisch A".

Alle Produktionsdaten stehen auf dem Zellencontroller für einige Schichten detailliert zur Verfügung. Es gilt aber das Grundkonzept, daß längerfristige Archivierungen dann von einem übergeordneten Leitstand durchgeführt werden sollen.

5.   Struktur des Zellencontrollers

Der Zellencontroller ist modular aus Programmpaketen aufgebaut. Das Basiselement dieses Aufbaus ist ein Echtzeit-Datenbanksystem. Hier legen einzelne Programmpakete intern benötigte Daten ab, andere Programmteile greifen über die Datenbank auf die für sie bestimmten Daten zu.

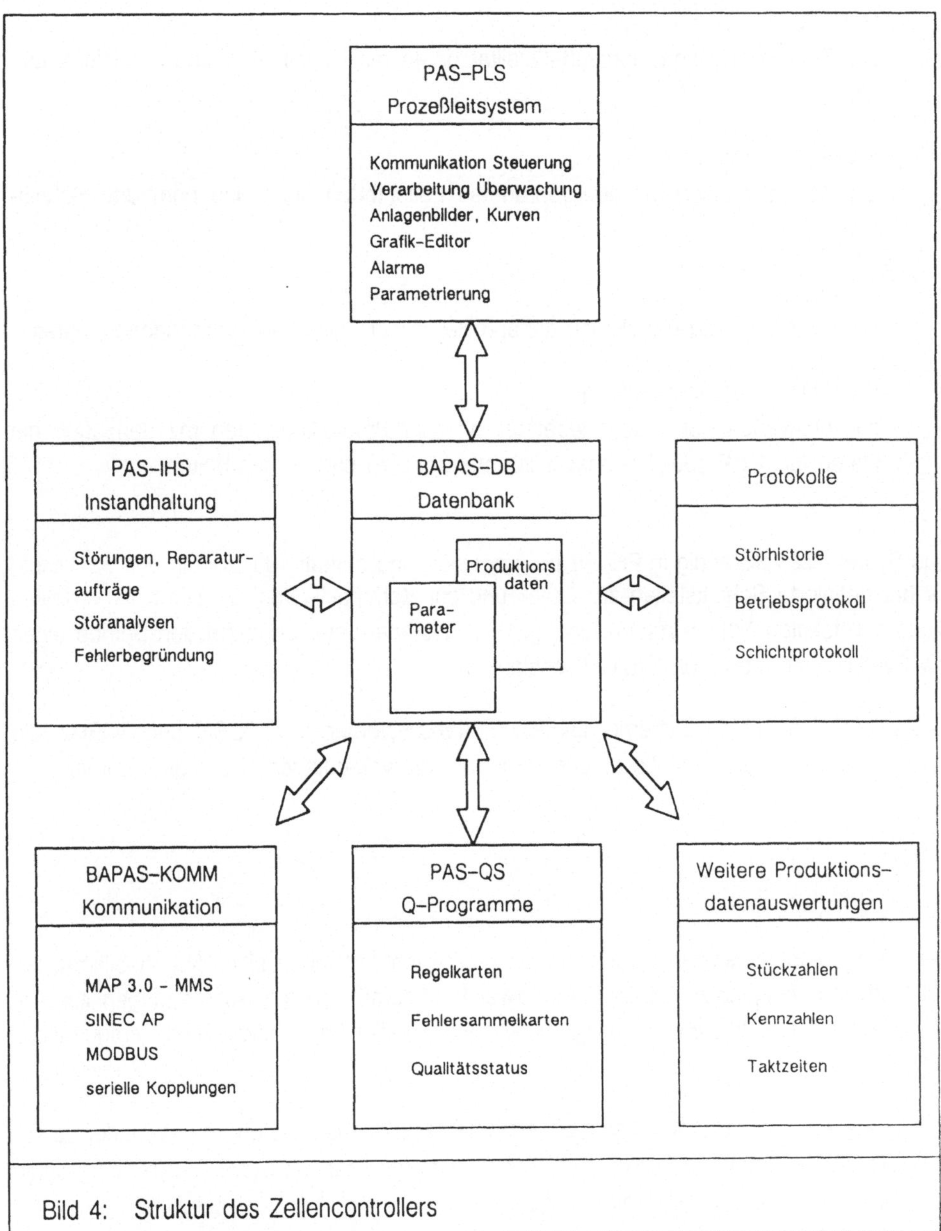

Bild 4:   Struktur des Zellencontrollers

o   PAS–PLS
   als Prozeßleitsystem liefert die Funktionen Datenaustausch mit den Steuerungen,
   Überwachen der Daten, Visualisierung in Anlagenbildern und Kurven, Alarmierung
   und Datenarchivierung

o   PAS–IHS
   zur Unterstützung der Instandhaltung mit Störmeldungen, Reparaturaufträgen und
   Begründung von Fehlern

o   PAS-QS

zur Bearbeitung von Inprozeß-Qualitätsdaten mit Regelkarten und Qualitätsstatus als Ergebnis

o   BAPAS-KOMM

zur Kommunikation mit übergeordneten Leitständen über eine normierte Schnittstelle

o   Protokolle

zur Erstellung von Protokollen anhand der in der Datenbank vorhandenen Daten

o   Produktionsdatenauswertung

zur Auswertung und Gegenüberstellung aller Produktionsdaten mit dem Ziel, die Daten für die Produktion besser zu interpretieren und zu bearbeiten

Das System ist vollständig in PEARL programmiert und enthält 193 Moduln und 32 Tasks. Es hat definierte Schnittstellen zur Grafik und zur Kommunikation. Es steht als verteiltes System in lokalen Netzen zur Verfügung. Dies gestattet unter anderem den Betrieb eines PC-Systems mit mehreren Grafik-Bildschirmen.

Es existieren lauffähige Systeme auf den Betriebssystemen VMS, OS/2 und SORIX. Alle Systeme werden aus derselben Mutterquelle auf die verschiedenen Systeme portiert.

6.   Ausblick

Der Zellencontroller wird in verschiedenen Produktionsbereichen einer Automobilfirma auf PC-Hardware eingesetzt. Es existieren zwei Konfigurationen mit Ausprägungen zur Anlagen- und Materialflußüberwachung zur Abdeckung aller angeschlossenen Fertigungszellen.

Ähnliche Aufgabenstellungen sind bei getakteten Fertigungsabläufen immer wieder zu finden. Deshalb konnten die Funktionen auch in anderen Branchen (z.B. Verpackungslinien oder Wandlerfertigung) eingesetzt werden.

Speziell in Verbindung mit einem Fertigungsleitstand wird der Zellencontroller zum Bestandteil einer modernen CIM-Struktur mit klarer Verteilung von Daten und Verantwortung. Es ist jedoch zu bemerken, daß ein Fertigungsleitstand sich in den Funktionen nicht grundlegend vom Zellencontroller unterscheidet sondern in dem Detaillierungsgrad der Daten und im zu überschauenden Zeitfenster.

Betriebsdatenerfassung in der CD Produktion*

V. Rasche, R. Siemens und W. Schepper

Fakultät für Physik, Universität Bielefeld
Universitätsstraße 25, D-4800 Bielefeld 1

Kurzfassung:

In dem hier vorgestellten Projekt werden die Maschinenbetriebsdaten der in einer CD Produktion
eingesetzten Maschinen erfaßt und ausgewertet. Da die verschiedenen Maschinen in der Regel über
unterschiedliche Maschinendaten-Schnittstellen verfügen, wurden zur Ankopplung an ein MBDE -
System kleine Rechner entwickelt, die untereinander innerhalb der Produktion vernetzt sind. Ihre
Aufgabe ist eine Protokollkonvertierung, Datenverdichtung und Vorauswertung. Die Rechner sind
mit einer leistungsfähigen CPU und einem Bus ausgestattet und können von einem übergeordneten
Entwicklungssystem frei programmiert werden. Die Hard- und Software- Flexibilität ermöglicht
leicht, neue Maschinen anzuschließen oder dieses System auch in anderen Fertigungen einzusetzen.
Durch das Protokollieren der Stillstandszeiten und der Fehlergründe erhält man objektive Aussagen
über den Zustand jeder einzelnen Maschine. Man kann die Maschinen untereinander vergleichen,
Wartungsarbeiten planen und rückwirkend auf die Maschineneinstellungen zugreifen. Da die Daten
netzwerkweit verfügbar sind, können leicht maschinenübergreifende Reports generiert werden.

Schlüsselworte

CIM, CAQ, BDE, MDE, Fabrik-Netzwerk, Prozeßvisualisierung, Qualitätssicherung, Maschinen-
Schnittstellen

1. Einleitung

Das hier vorgestellte Projekt befaßt sich mit der Betriebsdatenerfassung in einer CD ("compact-
disc") -Produktion. Da bei der Darstellung des Projekts immer wieder auf die verschiedenen Ferti-
gungsschritte Bezug genommen werden muß, soll hier kurz auf den komplexen Herstellungsprozeß
einer CD eingegangen werden. Die Vorlagen mit den Dateninformationen werden in der Regel vom
Besteller einer CD über Magnetband geliefert. Die Pits als Träger der Information werden über
einen Ar-Ion-Laser als Belichtungsquelle pulslängenmoduliert in eine Photoresistschicht auf einer
Glasplatte übertragen (ähnlich dem von der Chip-Herstellung bekannten Fotoätzverfahren), entwickelt und mit einer dünnen Silberschicht (1000 Å) überzogen. Galvanotechnisches Aufwachsen
einer Nickelschicht führt zur Vatermatrize, von der durch 2-maliges Umkopieren die Matrize für
die Spritzgußmaschine hergestellt wird. Im CD-Fertigungsprozeß sind ähnlich hohe Anforderung
an die einzuhaltenden Abmessungen zu stellen wie beim dynamischen Halbleiter-Speicher (Mega-
bitchip). Das kommt in folgenden Angaben zum Ausdruck: Tiefe der Pits $0.1\,\mu$, Spurabstand:

---

* unterstützt durch die Firma Sonopress/Bertelsmann, Gütersloh

1.6 μ, Länge der Pits: 0.8 - 3 μ. Die Matrize prägt in der Spritzgußmaschine den CD-Rohling aus durchsichtigem Kunststoff (Polycarbonat). Danach erfolgt auf einer Sputteranlage die Aluminium-Beschichtung (500 Å), die Schicht wird mit einer 0.1 mm Lackschicht geschützt. Danach wird die Lackschicht bedruckt und die jetzt fertige CD verpackt. Wie Abb. 1 zu entnehmen ist, finden alle Fertigungsprozesse bis einschließlich Lackierung wie bei der Chip-Herstellung im Reinstraum statt, erst die anschließende Prüfung der CD im Laser-Scanner, Druck und Verpackung der fertigen CD finden unter Normalbedingungen statt.

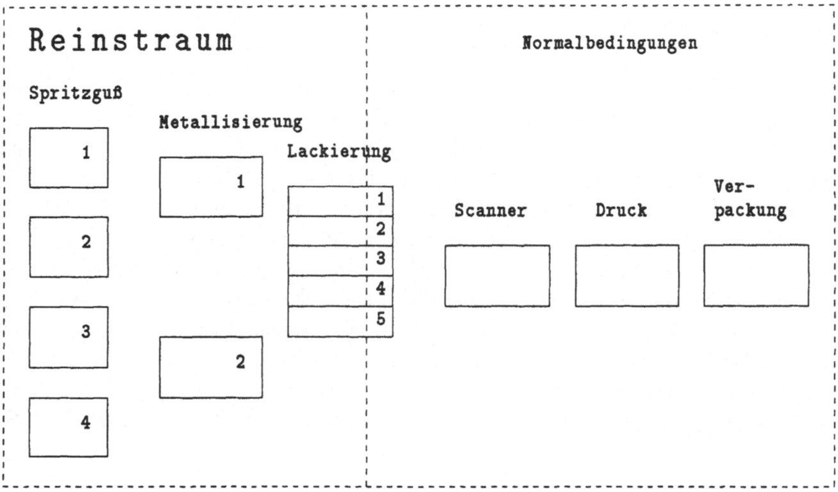

Abb. 1 Fertigungsstufen der CD

Ziel der Betriebsdatenerfassung ist eine Datenerhebung in Echtzeit, die jederzeit Angaben über den aktuellen Maschinenzustand liefert. Ein Ergebnis der Datenerhebung kann die Optimierung des Produktionsprozesses sein, die in einer Verringerung der Zykluszeit (Durchlaufzeit der CD innerhalb der einzelnen Maschine) und Stillstandszeit oder auch Erhöhung der Verfügbarkeit für jede einzelne Maschine zum Ausdruck kommt. Ein anderes Ergebnis kann darin liegen, daß die mit der Datenauswertung mögliche Fehleranalyse (Häufung von Fehlern, Rückverfolgbarkeit von Fehlermeldungen) gezielte Servicearbeiten der betroffenen Maschine ermöglicht. Insgesamt kann dadurch der Wartungsaufwand reduziert werden, die vorzeitige Korrektur von Fehlern führt zu einer Qualitätsverbesserung. Produktionsmaschinen weisen eine große Anzahl von Einstellparametern auf (über 100 Parameter bei der Spritzgußmaschine z. B.). Computerunterstützte Optimierungsverfahren müssen zur Qualitätsverbesserung genutzt werden. Ohne Computerhilfe kann eine Optimierung durch systematische Veränderung einer Vielzahl von Einstellparametern nicht zum Ziel führen. Die Betriebsdatenerfassung kann dazu führen, daß die Fertigungsmaschinen auf Grund vorliegender quantitativer Angaben beurteilt werden können. Das ist sehr wichtig als Grundlage für Kaufentscheidungen (Vergleich einer Blockfertigungsstraße und eines Monoliners z. B.). Es ist auch möglich, die in den Kaufverträgen festgelegten Leistungsdaten der Maschinen (garantierte Verfügbarkeitsangabe von 90% z. B.) zu kontrollieren.

2. Maschinen-Betriebsdaten-Erfassung - MBDE

Mit diesem System sollen die Daten der Maschinen in der CD-Produktion erfaßt werden. Durch eine genaue Protokollierung der Einstellwerte, der Stillstandszeiten und der Stillstandsgründe der Maschinen soll erreicht werden, daß man den Erfolg von Verbesserungen an den Maschinen sofort und objektiv beurteilen kann, daß eingetretener Verschleiß an den Maschinen festgestellt werden kann und frühzeitig Reparaturarbeiten geplant werden können.

Die Daten, die in dieses System eingehen, kann man grob in zwei Gruppen unterteilen :

1. generell für alle Maschinen :

    - Stillstandszeiten
    - Stillstandsgründe
    - Häufigkeit der auftretenden Fehlermeldungen
    - Laufzeiten
    - Stückzahlen
    - Verfügbarkeit

2. maschinenabhängig :

    - Einstellwerte
    - Steuerungsprogramme

Diese Daten sollen dabei im "online"-Zugriff sein, um jederzeit eine komplette Übersicht über die gesamte Produktion zu haben. Desweiteren sollten sie aber auch in einer Datenbank abgespeichert werden, um daraus Tages- und Wochenreports generieren zu können.

Der Anschluß der Maschinen ist über geeignete Schnittstellen realisiert und erfolgt vollautomatisch. Maschinen, an denen keine Rechner-Schnittstelle vorhanden war, wurden damit nachgerüstet. Eine vollautomatische Datenerfassung vermindert den Anfall von fehlerhaften Daten und kostet keine zusätzliche Arbeitszeit für eine Terminaleingabe.

Tabelle 1 Schnittstellen und Protokolle :

Maschine	Hardware-schnittstelle	Protokoll	Reaktion	Fehler-kontrolle
Spritzguß	V.24	Hersteller spezifisch	Antwort auf Request	Parity
Metallisierung	SPS	Ausgabebyte	Meldung ohne Request	-
Metallisierung	V.24	Hersteller spezifisch	Meldung auf Request	-
Metallisierung	V.24	Bedienerterminal	Meldung ohne Request	-
Lackierung	V.24	SECS	beides programmierbar	Checksumme
Scanner	SPS	Ausgabebyte	Meldung ohne Request	-
Druck	SPS	Ausgabebyte	Meldung ohne Request	-
Druck	SPS	Ausgabebyte	Meldung ohne Request	-
	V.24	Terminal an OS-9	Antwort auf Request	Parity möglich

Tabelle 1 gibt eine übersicht der Schnittstellen und Protokolle der verwendeten Maschinen. Zur Steuerung der Maschinen werden unterschiedlich intelligente Steuerungen verwendet. Dies reicht von einer SPS Steuerung, deren Speicher mit dem Steuerprogramm bereits fast voll ist, bis hin zu einer Mehrprozessor-Steuerung, die für die externe Kommunikation einen eigenen Prozessor besitzt. Dadurch kann diese Steuerung ein aufwendiges, aber sehr mächtiges Protokoll (SECS [1]) unterstützen. Ein besonders wichtiger Punkt sind die Kosten, die das Nachrüsten der Maschinen mit entsprechenden Schnittstellen oder das Nachprogrammieren entsprechender Schnittstellen durch den Maschinenhersteller verursacht. Für die SPS Steuerung der Fa. Siemens gibt es zur Ankopplung externer Rechner einen Kommunikationsprozessor. Dieser ist aber so teuer, daß allein die Kosten für dessen Einsatz den Finanzrahmen dieses Projektes übersteigt. Diese Steuerungen wurden deshalb über ein Ausgabebyte der SPS angeschlossen.

Tabelle 2 Datenmengen und Zykluszeiten :

Maschine	Hardware-schnittstelle	Meldungen / Tag	Zykluszeit [s]	Programmierung bei sonopress möglich
Spritzguß	V.24	70000	5.0 - 99.0	nein
Metallisierung	SPS	30000	2.0 - 5.0	ja
Metallisierung	V.24	3000	1.0 - 3.0	nein
Metallisierung	V.24	1000	1.0 - 3.0	nein
Lackierung	V.24	60000	5.0 - 40.0	nein
Scanner	SPS	60000	1.0 - 4.0	ja
Druck	SPS	40000	2.0 - 5.0	ja
	V.24	100	2.0 - 5.0	ja

In Tabelle 2 sind die an den Schnittstellen anfallenden Datenmengen und die Zykluszeiten der Maschinen angegeben. Die Meldungen/Tag geben dabei eine obere Grenze der anfallenden Meldungen. Als untere Zykluszeit-Grenze ist ungefähr 50% der aktuellen Zykluszeit angegeben. Damit sind auch zukünftige Verbesserungen an den Maschinen berücksichtigt. Die anfallenden Meldungen und die Zykluszeit bestimmen die Reaktionszeit des abfragenden Rechners. Aus der Tabelle ist ersichtlich, daß diese im Sekundenbereich liegen muß. Dies ist für einen Mikroprozessor absolut unkritisch. Die Meldungen / Tag machen deutlich, welche großen Datenmengen hier anfallen. Wichtig ist auch, ob eine Steuerung in der Firma programmiert werden kann, oder ob dies vom Maschinenhersteller geschehen muß. Im letzteren Fall ist der große Aufwand mit erheblichen Kosten verbunden. Umgekehrt muß man bei den im Hause programmierbaren Steuerungen damit rechnen, daß die entsprechenden Programme auf dem UMaKS angepaßt werden müssen.

Wie man aus der Tabelle sieht, haben die meisten Maschinen unterschiedliche Schnittstellen und benutzen zur externen Datenkommunikation die unterschiedlichsten Protokolle. Eine Vereinheitlichung der Schnittstellen und der zugehörigen Protokolle durch den Maschinenhersteller ist in einigen Fällen nicht möglich. Eine Änderung der Maschinen-Steuerprogramme vom Hersteller ist in jedem Fall sehr teuer.

Um an die Daten aller Maschinen zu gelangen, ist es günstiger, einen kleinen Koppelrechner zwischen Maschine und Datenerfassungs-System zu setzen. An diesen müssen folgende Anforderungen gestellt werden :

- echtzeitfähig
- hohe Verfügbarkeit
- Schnittstellen zu allen Maschinen
- Protokollkonvertierung zu einem einheitlichen Protokoll
- netzwerkfähig
- anpaßbar an jetzt noch unbekannte Maschinen, deren Schnittstellen und Protokolle
- für den Einsatz in der Produktion geeignet

Er soll folgende Aufgaben übernehmen :

- Datenerfassung
- Datenverdichtung
- Schichtauswertung
- Datenpufferung

Da der benötigte Rechner insbesondere zu allen vorhandenen Maschinen Schnittstellen haben muß, andererseits aber auch nicht zu teuer sein sollte, wurde solch ein Rechner neu entwickelt. Er erhielt die Abkürzung : **UMaKS** = (Universelles **Ma**schinen **K**ommunikations **S**ystem).

2.1 Hardware

Die Schwerpunkte bei der Entwicklung des UMaKS lagen in der Bereitstellung möglichst hoher Verfügbarkeit, Echtzeitfähigkeit und einer großen Flexibilität bezüglich späterer Erweiterungen. Ferner muß der Rechner eine genügend große Rechenleistung zur Verfügung stellen, um den Anforderungen der Datenerfassung und Verarbeitung (siehe Kap 2.) gerecht zu werden. Bereits bei der Konzeption des Rechners stand fest, daß er an neue Maschinen oder Protokolle angepaßt werden muß, die zum Zeitpunkt der Entwicklung unbekannt waren. Da diese Anpassungen sowohl Hard-, wie auch Software- änderungen umfassen, war es sinnvoll, bei der Wahl des Prozessors einen gebräuchlichen Typ mit einer komfortablen Entwicklungsumgebung zu wählen.

Um den Anforderungen der Datenerfassung und Verarbeitung (Bereitstellung großer linear adressierbarer Speicherbereiche, Multiuser/Multitasking, Netzwerk ...) gerecht zu werden, wurde als Prozessor der MC68000 von Motorola eingesetzt. Für ihn existiert mit dem Betriebssystem OS-9 eine komfortable Entwicklungsumgebung und mit dem VMEbus ein weit verbreitetes Bus-System.

Die Ausmaße des UMaKS mußten möglichst gering gehalten werden, um den Einsatz in vorhandenen Schaltschränken zu ermöglichen. Aus diesem Grund wurde als Platinengröße eine einfache Europakarte gewählt. Um allen Schnittstellen-Anforderungen der Datenerfassung (siehe Kap. 2.) gerecht zu werden, mußte von einer Einplatinenlösung abgesehen werden.

Bei der Realisierung des UMaKS wurden 3 Platinen entwickelt, die über eine "VMEbus-Backplane" verbunden wurden :

- **CPU – Platine**

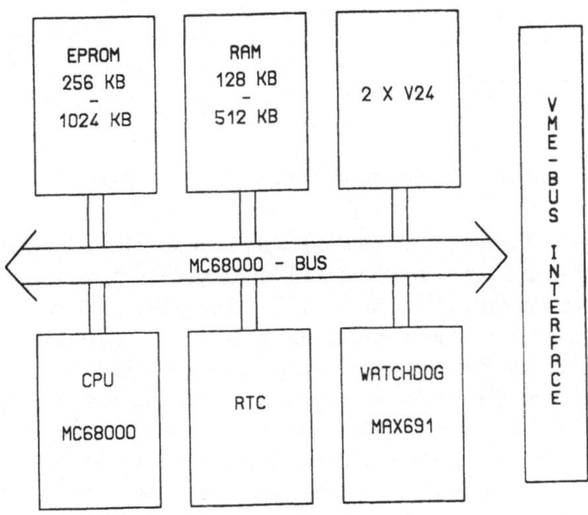

Abb. 2 Blockschaltbild CPU – Platine

Die CPU – Platine (siehe Abb. 2) stellt einen kompletten Rechner auf Basis des MC68000 dar. Um der Karte Echtzeiteigenschaften zu ermöglichen, sind eine Echtzeituhr (RTC72421), ein Watchdog (MAX 691) und ein batteriegepuffertes RAM eingesetzt. Der MAX 691 übernimmt zusätzlich die Spannungsüberwachung des UMaKS. Die Steckplätze des RAMs sind so ausgelegt, daß vier 32 kByte oder vier 128 kByte RAMs eingesetzt werden können. Zur Aufnahme des Betriebssystems befinden sich zwei EPROM-Steckplätze auf der CPU – Karte, die wahlweise 128 kByte, 256 kByte oder 512 kByte EPROMs aufnehmen können.

Zur Kommunikation wird ein MC68681 eingesetzt, der zwei V.24 Schnittstellen zur Verfügung stellt. Die Steuerung der Treiber, die Erzeugung des $\overline{\text{DTACK}}$ – Signals, das Interrupt – Handling und die Adress – Dekodierung wird über drei GALs realisiert. Im Normalfall wird der Rechner innerhalb der Produktion ohne Terminal betrieben. Zur lokalen Diagnose werden vier LEDs eingesetzt, die zum einen Informationen über den Hardwarezustand des Rechners liefern, aber auch vom Betriebssystem genutzt werden, um den Softwarezustand anzuzeigen.

- **Input/Output – Platine**

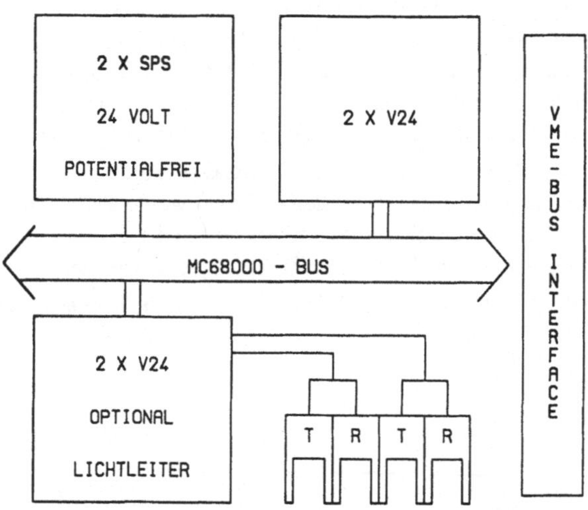

Abb. 3 Blockschaltbild Input/Output – Platine

Die Input/Output – Platine (siehe Abb. 3) stellt die benötigten Schnittstellen zur Erfassung der digitalen Daten und zur Ankopplung des UMaKS an ein Netzwerk zur Verfügung. Zur Erfassung der Daten von einer SPS wird ein MC68230 eingesetzt, der zwei 8 Bit breite parallele Schnittstellen besitzt, für die je zwei unabhängige Interrupt – Eingänge existieren. Zur Trennung der Potentiale von Rechner und Maschine sind die Eingänge über Optokoppler potentialfrei gehalten. Daten, die von den Maschinen über V.24 Schnittstellen bereitgestellt werden, können über zwei V.24 Schnittstellen erfaßt werden, die optional "Hardware-Handshake"- Signale (CTS/RTS) unterstützen können. Zur Einbindung des UMaKS in ein Netzwerk sind auf der Karte zwei V.24 Schnittstellen vorhanden. Diese sind für den Einsatz von TTL - - Lichtleiter – Umsetzern ausgelegt. Es können optional Umsetzer für Kunststoffaser oder Glasfaser eingesetzt werden. Die Ankopplung der UMaKS Rechner an ein Netzwerk über Lichtleiter empfiehlt sich, da die Lichtleiter unempfindlich gegenüber den hohen Störpegeln innerhalb einer Fabrikumgebung sind, und gleichzeitig eine galvanische Entkopplung der Rechner erreicht wird. Zusätzlich befindet sich auf der Platine ein MC68153, der das Interrupt – Handling für den VMEbus (daisy chained , vectored Interrupts) übernimmt. Die Steuerung der Treiber und die Adress – Dekodierung übernimmt ein GAL.

- **ADC – Platine**

Die ADC – Platine (siehe Abb. 4) wird zur Erfassung analoger Signale eingesetzt. Durch die Verwendung des DSP56ADC16 von Motorola können Signale bis zu 200 kHz bei 12 Bit

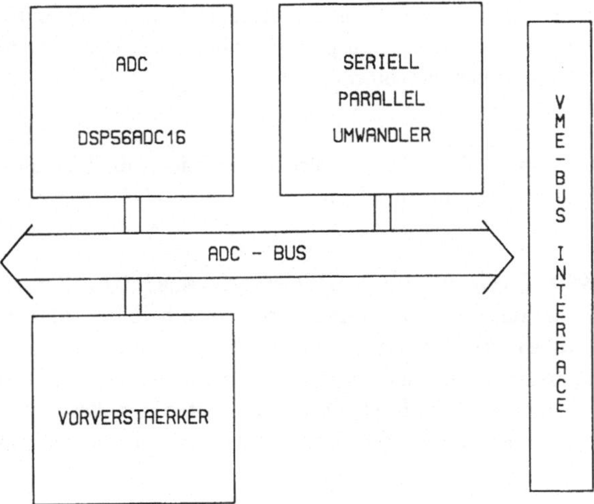

Abb. 4 Blockschaltbild ADC – Platine

Auflösung und bis 50 kHz bei 16 Bit Auflösung erfaßt werden. Der DSP56ADC16 besitzt ein serielles Interface. Zur Wandlung des seriellen Signals in ein 16 Bit Wort werden zwei 74LS595 eingesetzt. Die Adress-Dekodierung und die Erzeugung des $\overline{\text{DTACK}}$ Signals erfolgt durch ein GAL.

2.2 Software

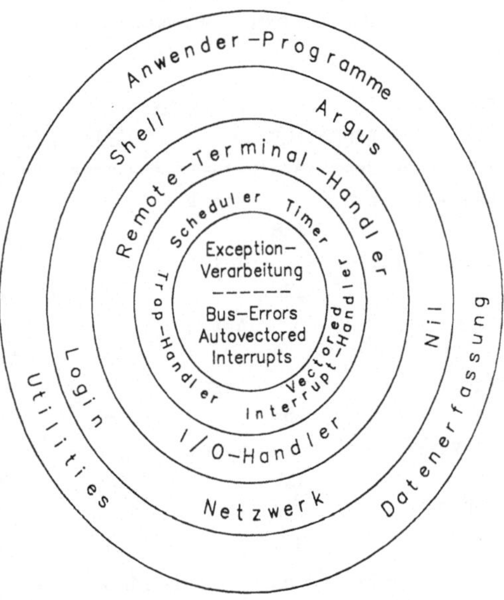

Abb. 5 Aufbau Betriebssystem UMaKS

Die Gliederung des Betriebssystems des UMaKS in Kernal (Nukleus), Schnittstellen-Software und Anwendermodule (siehe Abb. 5) vereinfacht ganz erheblich die spätere Erweiterung der Software, wie z.B. die Unterstützung neuer Maschinen.

2.2.1 Kernal

Der Kernal von ROPS_2 ist in Assembler geschrieben und stellt die fundamentalen Routinen des Betriebssystems zur Verfügung. Hierzu zählen :

- Installation :

  Wird der UMaKS erstmalig gebootet, besitzt er noch keine Informationen über die angeschlossenen Maschinen, zu installierende Prozesse usw.. Innerhalb des Bootvorganges hat der Anwender die Möglichkeit, den Rechner für seine Aufgabe zu konfigurieren. Dieses geschieht interaktiv durch Eingabe der zu installierenden Prozesse und der zu installierenden Schnittstellen-Treiber. Nach Eingabe wird die Konfiguration im batteriegepufferten RAM abgelegt, so daß sich der Rechner nach einem erneuten Bootvorgang selbstständig konfigurieren kann.

- Initialisierung :

  In der Initialsierungsphase wird der UMaKS in einen wohldefinierten Zustand versetzt. Insbesondere wird dabei sichergestellt, daß die Datenerfassung automatisch gestartet wird. Durch den Einsatz des batteriegepufferten RAMs werden bei einer Reinitialisierung des UMaKS bereits erfaßte Daten nicht gelöscht.

- Prozeßumschaltung

  ROPS_2 unterstützt bis zu acht Prozesse. Die Verteilung der Rechenzeit erfolgt an Hand der Priorität der einzelnen Prozesse. Besitzen mehrere Prozesse die gleiche Priorität, wird die Rechenzeit zwischen ihnen nach *Round Robin* [2] aufgeteilt. Um Prozessen geringerer Priorität Rechenzeit zu verschaffen, ist mit jedem Prozeß ein Status verbunden. An einem Prozeß wird nur Rechenzeit vergeben, wenn er den Status *aktiv* besitzt. Im Zustand *blocked* wird er bei der Verteilung der Rechenzeit nicht berücksichtigt. Ein Prozeß erhält den Status *blocked*, wenn er auf die Beendigung einer IO – Operation wartet, explizit von der Software in den Status "sleeping" gelegt wird oder bereits suspendiert ist.

```
    JOBID   JOB-NAME                 PROGRAMM              STATUS
    TICKS                 PRIORITAET          DEFIN DEFOUT
    ===========================================================
        0   system login            shell                 ru
    248CD                     128              0     0
        1   remote login            login                 ak
    41A2D                     128              2     2
        2   nil                     nil                   ak
    0                           0             -1    -1
        3   argus                   argus                 sl
    94                        255             -1    -1
        4   netzwerk                sonet                 wti
    10A3E                     128              3     3
        5   MBDE-SP                 sp                    sl
    32AE                      200              5     5
```

Abb. 6 Beispiel einer typischen Installation

Durch den Befehl *show_sys* innerhalb der Shell, wird eine Übersicht der installierten Prozesse ausgegeben (siehe Abb. 6). Neben der Prozeßidentifikations-Nummer, dem Prozeßnamen und dem aktiven Programm wird der Status (*ru* running, *ak* active, *wti* waiting till interrupt, sleeping, *st* stopped), sowie verbrauchte CPU – Zeit, Priorität und I/O – Schnittstellen angezeigt.

- Exception-Verarbeitung

  Tritt unter ROPS_2 eine Exception auf Grund eines Fehlers (Buserror, Adresserror, Divide by zero ...) auf, so wird der Fehlergrund und der Zeitpunkt seines Auftretens in eine Tabelle eingetragen und der Rechner anschließend neu gebootet. Bei einem "*autovectored*"-Interrupt erfolgt der Sprung in die Interrupt-Serviceroutine direkt über den entsprechenden Vektor. Die Abarbeitung eines "*vectored*"-Interrupts erfolgt über einen Sprung in Routinen des Betriebssystems, so daß die Möglichkeit einer Parameterübergabe an die Interruptroutine gegeben ist.

- System-Timer

  ROPS_2 besitzt einen System-Timer mit einer Auflösung von 20 ms. An Hand dieses Timers erfolgt die Prozeßumschaltung und das Rücksetzen des Watchdogs. Zur Vermeidung eines Auseinanderlaufens des System-Timers und der Echtzeituhr, wird der System-Timer um 0.00 Uhr mit der Echtzeituhr synchronisiert.

- Trap – Handler

  Standardroutinen werden vom Kernal über Trap Handler zur Verfügung gestellt. Zur Zeit existieren Trap Handler für :
  - Systemroutinen (Speicherverwaltung, Job – Handling, Job – Kommunikation ...)
  - Input-/Output Routinen
  - Utilities (Echtzeituhr abfragen und setzen, Datenkonvertierung ...)
  - Routinen für das Queue – Handling

- Queue – Handler

  Rops_2 unterstützt bis zu vier Queues. Vom Betriebssystem wird eine Queue als Timerqueue verwendet. Von den restlichen drei Queues wird zur Zeit eine als Netzwerkqueue eingesetzt.

- Speicherverwaltung

  Der Speicher wird dynamisch verwaltet. Ein Speicherblock kann *permanent* oder *nonpermanent* angefordert werden. Ein *permanent* angeforderter Speicherblock wird im Gegensatz zu einem *nonpermanent* Speicherblock bei einem Bootvorgang des Rechners nicht freigegeben, so daß eine ins RAM geladene Software auch nach einem Bootvorgang noch zur Verfügung stehen kann.

- System Debugger

  Als Debugging-Tool ist innerhalb des Kernals ein Debugger implementiert, durch den jeder Speicherbereich des UMaKS angezeigt und editiert werden kann. Daneben besteht die Möglichkeit, einzelne Programme im "*single step mode*" abzuarbeiten.

2.2.2 Schnittstellen-Software

Alle IO – Operationen werden unter ROPS_2 über Schnittstellen-Handler abgearbeitet. Diese verfügen im allgemeinen über einen Ringpuffer, dessen Daten per Interrupt abgearbeitet werden, so daß ein Prozeß nicht auf die Beendigung der Kommunikation warten muß, sondern seine Daten lediglich in den Ringpuffer einträgt. Ist der Ringpuffer voll, und kann ein Prozeß eine IO – Operation nicht abschließen, wird er vom System in den Status *sleeping* gesetzt, damit er keine Rechenzeit mehr verbraucht. Hat die Schnittstellen-Software die IO – Operation beendet, weckt sie den schlafenden Prozeß wieder auf.

2.2.3 Anwendermodule

Anwendungs-Software wird unter ROPS_2 in Form von OS-9 – Modulen realisiert, so daß dem Anwender eine komfortable Entwicklungsumgebung zur Verfügung steht. Bibliotheken für den Zugriff auf die Trap Handler existieren zur Zeit für Assembler und C.

Durch die Aufteilung der ("reentrant-") Programm-Module in Variablen- und Programm-Code-Bereiche braucht ein Modul, das von mehreren Benutzern gleichzeitig benutzt werden soll, nur einmal im Speicher des Rechners vorhanden sein.

Da sowohl der Kernal wie auch die einzelnen Programm-Module unabhängig von der Adresse lauffähig sind, kann das komplette Betriebssystem einschließlich häufig benutzter Programm-Module in einem EPROM abgelegt werden. Software für Spezialanwendungen kann über das Netzwerk in den RAM – Bereich des Rechners geladen werden.

Einige wichtige, existierende Programmodule :

ARGUS	zuständig für Datenerfassungs-überwachung
INITJOB	Reinitialisierung eines Prozeß
LOADMOD	lädt OS-9 Programm-Module über Netzwerk
LOGIN	übernimmt Kontrolle der Zugangsberechtigung
MBDE	lokale Datenauswertung
SHELL	Benutzeroberfläche für ROPS_2
SONET	Netzwerkserver
TMODE	Konfiguration von V.24 Schnittstellen
UNLOADMOD	Entfernen eines Moduls aus dem RAM – Bereich

Zusätzlich existieren diverse Module zur Datenerfassung, Protokoll- und Schnittstellen – Tester ...

2.2.4 Reaktionszeiten und maximale Datenraten

Die Reaktionszeiten und maximalen Datenraten hängen von der Einbindung der Datenerfassung ab. Um maximale Geschwindigkeiten zu erzielen, ist der Einsatz spezieller Hardware (z.B.: ADC mit FIFO ...) denkbar. Auf Software - Ebene hängen die Reaktionszeiten von der Implementierung ab. Am schnellsten sind Interrupt-Routinen, die direkt angesprungen werden. Eine normale Interrupt-Einbindung mit Einsprung ins Betriebssystem ist etwas langsamer, dafür aber flexibler. Ist die Datenerfassung über einen Prozeß realisiert, muß berücksichtigt werden, daß durch das Timesharing bis zu 20 ms verloren gehen können, bis der Datenerfassungs-Job Rechenzeit bekommt.

Einbindung	Reaktionszeit	maximale Datenrate
Hardware	hardwareabhängig	hardwareabhängig
Interrupt indirekt	ca. $40\ \mu s$	10.9 kHz
Prozeß	≥ 20 ms	≤ 50 Hz

Die Reaktionszeit und maximale Datenrate wurden bei einer Interruptroutine zur Meßdaten-Erfassung über eine SPS–Schnittstelle bestimmt.

Um die Differenz in der Reaktionszeit zwischen direkter und indirekter Ankopplung zu vergleichen, wurde eine Routine verwendet, die nur den Interrupt – Request auf dem Chip zurücksetzt. Bei direkter Einbindung wird eine Reaktionszeit von ca. 11 μs und bei indirekter Einbindung von ca. 40 μs erreicht. Die Reaktionszeiten liegen somit in der gleichen Größenordnung - wie für RTOS auf einem ATARI ST [3,4] angegeben.

2.2.5 Datenerfassung

Durch die Zwischenspeicherung der Maschinendaten auf den lokalen Rechnern wird der Host-Rechner von Echtzeitaufgaben entbunden. Die lokalen Rechner nehmen eine Daten-Kompression vor, indem sie schichtbezogen die anfallenden Maschinenmeldungen in einer Statistik ablegen. In der Statistik werden für jede mögliche Fehlermeldung Anzahl und Zeit des Auftretens vermerkt. Die Datenerfassungs – Software ist so ausgelegt, daß auf dem lokalen Rechner die Meldungen aus bis zu 8 Schichten gepuffert werden können. Zusätzlich zu den Statistiken werden die letzten Meldungen (z. B. 50) in einem Ringpuffer zwischengespeichert. So ist die Möglichkeit gegeben, alle Meldungen an den Host-Rechner durchzureichen.

Eine Auswertung der erfaßten Daten kann bereits auf den lokalen Rechnern erfolgen. So hat der Anwender die Möglichkeit, auch direkt an der Maschine Informationen über den Produktionsablauf zu erhalten.

Die Maschinenmeldungen werden im Normalfall über Interrupts erfaßt. Durch die geringen Reaktionszeiten ist es somit möglich, den Meldungen eine Uhrzeit zuzuordnen. Da die Frequenz der innerhalb der CD–Produktion anfallenden Meldungen 10 Hz nicht übersteigt, ist sichergestellt, daß keine Meldungen verloren gehen (siehe Tabelle 2).

2.2.6 Netzwerk

Das Netzwerk ist hauptsächlich zur Entsorgung der lokalen Rechner gedacht. Pro UMaKS fallen pro Tag ca. 6 kByte an Daten an. In der Produktion sind maximal 10 Rechner je Netzwerkstrang vorhanden, so daß die Datenmenge pro Netzwerkstrang pro Tag ca. 60 kByte beträgt. Die Datenrate auf dem Netzwerk kann also sehr gering sein. In unserem Projekt ist zur Zeit eine Baudrate von 19200 Baud eingestellt. Da die lokalen Rechner in der Lage sind, die Daten von 7 Schichten zu speichern, werden an das Netzwerk keine hohen Anforderungen an die Ausfallsicherheit gestellt. Bei der jetzigen Installation sind die lokalen Rechner über eine lineare Kette an den Host – Rechner angebunden.

Der Netzwerkserver auf den UMaKS-Rechnern ist ein Programm-Modul. Die Kommunikation mit benachbarten Rechnern erfolgt über V.24 Schittstellen. Die Software unterstützt 4 Schichten des OSI–Referenzmodells :

Physical Layer (V.24)

Link Layer

Routing Layer

Session Layer

Es besteht die Möglichkeit, über das Netzwerk zu jedem UMaKS bis zu 4 *"Remote Sessions"* aufzubauen.

Es ist bereits erwähnt worden, daß der UMaKS als Konverter für das umfangreiche SECS-Protokoll [1] eingesetzt wird, eine Anpassung an das gebräuchliche DDCMP - Protokoll [5] von DEC konnte inzwischen ebenfalls demonstriert werden, eine Anpassung an andere, in Fertigungen eingesetzte Fieldbus-Systeme ist wegen des modularen Aufbaus der Software leicht möglich.

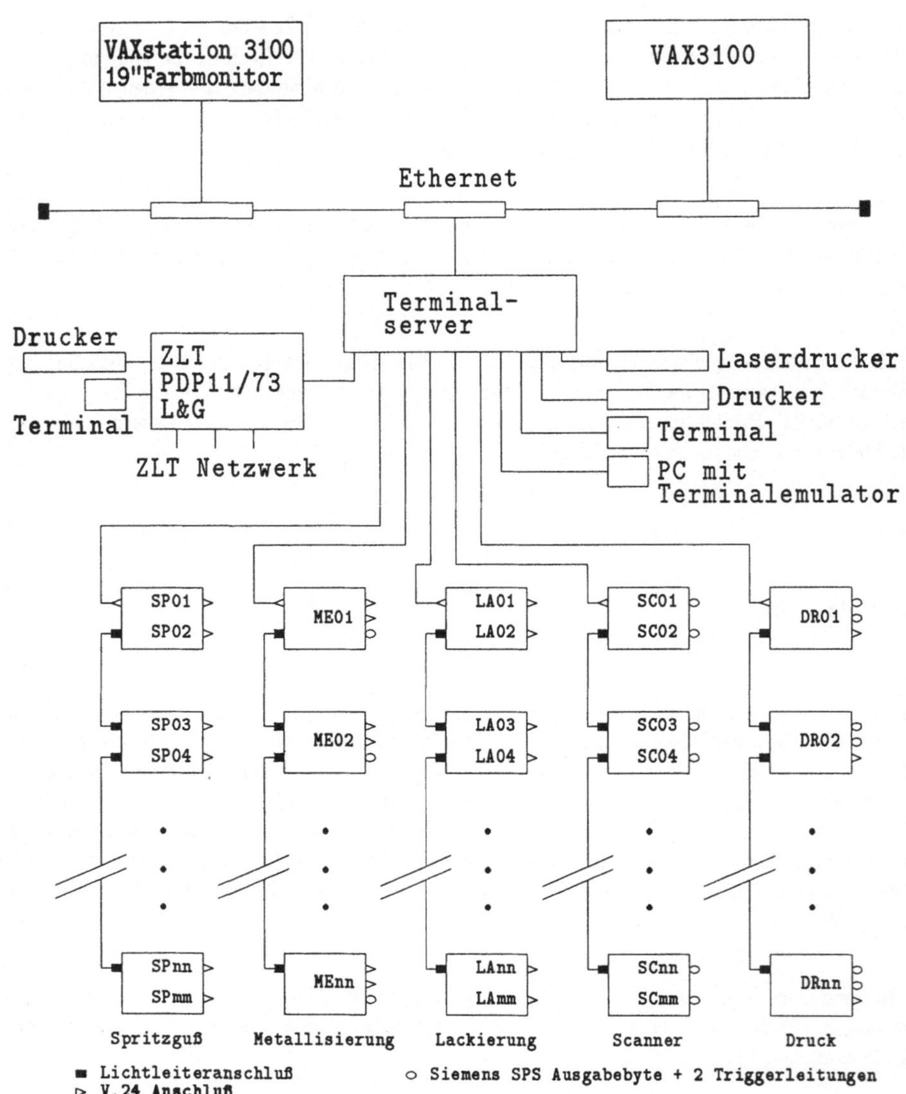

Abb. 7 Vernetzung der UMaKS

Die UMaKS - Rechner sind untereinander über Lichtleiter vernetzt. Die Anordnung der UMaKS innerhalb der Lichtleiterstränge ist beliebig und jederzeit änderbar. Die UMaKS Rechner übernehmen die Datenerfassung, machen eine schichtabhängige Auswertung und stellen diese Daten netzwerkweit zur Verfügung. Durch das Abspeichern mehrerer Schichten ermöglichen sie ein Ausschalten des übergeordneten Rechners für einige Tage, ohne daß ein Datenverlust auftritt. Der Anschluß an übergeordnete Rechner geschieht über einen Terminalserver. Dies macht das UMaKS Netzwerk physikalisch unabhängig von dem übergeordneten Rechner.

Der Terminalserver ist über Ethernet mit zwei VAXen verbunden. Eine VAXstation 3100 und eine VAX 3100. Die VAXstation 3100 ist als komfortabler "multiwindow-" und grafikfähiger Enwicklungsplatz gedacht, um das gesamte System weiterzuentwickeln. Auf der VAX 3100 liegt die Datenbank und die Netzwerksoftware für die UMaKS Rechner. Der Anschluß des UMaKS - Netzes über einen Terminalserver macht es möglich, die Netzwerksoftware für die UMaKS ohne Hardwareänderungen auf die andere VAX umzuschalten. Die Software dazu ist auf beiden Rechnern vorhanden, aber jeweils nur auf einem in Betrieb, die Aktivierung auf der anderen VAX ist jedoch jederzeit möglich.

3. Datenanalyse

Das Eintragen der Schichtauswertungen und der Maschineneinstellungen in die Datenbanken geschieht in zwei Schritten :

1. Übertragen der Daten eines UMaKS in eine Schichtdatei auf der VAX,

2. Eintragen der Datei in die Datenbank.

Die Schichtdateien werden auf Magnetband gesichert. Dies ermöglicht bei einem Crash des Systems, auch alte Backups der Datenbank mit Hilfe der Schichtdateien wieder auf den aktuellen Stand zu bringen. Dies Verfahren benötigt wesentlich weniger Speicherplatz und Zeit als das regelmäßige Sichern der **relativ großen Datenbankdateien**.

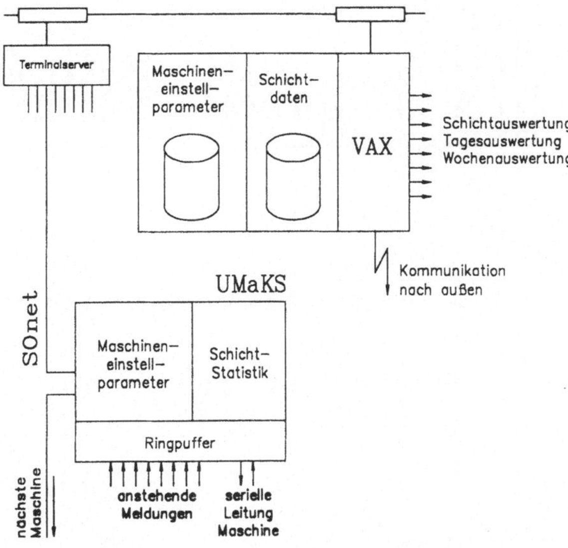

Abb. 8 Daten im MBDE-SYSTEM

Abbildung 8 zeigt das Datenmanagment in dem MBDE System. Auf den UMaKS Rechnern werden die anlaufenden Meldungen in einem Ringpuffer gespeichert. Dadurch kann der Maschinenbediener dem Servicepersonal nachweisen, daß die Maschine gerade wegen eines bestimmten Fehlers stehen geblieben ist. Dies reduziert das innerbetriebliche Konfliktpotential. Häufig läuft eine Maschine fehlerfrei, solange der Servicetechniker anwesend ist, und bleibt stehen, wenn er wieder weg ist. Die Größe des Ringpuffers und die Häufigkeit der auftretenden Meldungen bestimmt die Zeit für die Rückverfolgung der Meldungen.

Die Meldungen werden auf dem UMaKS schichtweise verdichtet. Auf dem UMaKS können die Auswertungen der letzten 8 Schichten angesehen werden. Außerdem wird auf dem UMaKS die Maschineneinstellung zwischengespeichert.

MBDE-SYSTEM V0.4   Anlage ME-01-IL

Daten   von : Mi 19.10.88   14:00:00   KW42/2
         bis : Mi 19.10.88   22:00:00

Stillstandszeit : 01:02:09
Laufzeit   : 06:57:51
Stückzahl   :   8383   mittlere Zykluszeit 2.99s

Meldung Nr	Text	Stillst. Gesamtzeit [St:Min:Sek]	Häufigkeit
25	SCHUTZABDECKUNG GEÖFFNET	00:18:51	79
17	INNENMASKE LAGENKONTROLLE AUSGABE...	00:09:24	60
38	FREIGABE CD AUSGABE FEHLT	00:04:47	61
39	FREIGABE AN BALZERS ERTEILT	00:04:45	9
57	NEUE ORDER AUSGABESTATION	00:03:18	27
5	STÖRUNG AUSGABETISCH	00:03:10	20
28	ANLAGE BETRIEBSBEREIT	00:03:08	104
18	CD AUS CARRIER NICHT ENTNOMMEN	00:03:02	42
60	CD-EINGABE MAGAZIN LEER	00:01:58	23
9	STÖRUNG CD-EINGEBER	00:01:52	19
4	STÖRUNG EINGABETISCH	00:01:35	9
33	FREIGABE CD EINGABE FEHLT	00:01:33	35
15	AUßENMASKE LAGENKONTROLLE	00:00:55	10
16	INNENMASKE LAGENKONTROLLE EINGABE...	00:00:54	6
61	CD-AUSGABE MAGAZIN VOLL	00:00:46	14
62	NEUE ORDER EINGABESTATION	00:00:44	19
3	AUSGABETISCH NICHT IN POSITION	00:00:23	8
52	INNENMASKE EINGEBER MAGAZIN LEER	00:00:20	3
13	STÖRUNG INNENMASKE AUSGEBER	00:00:19	2
14	STÖRUNG CD-AUSGEBER	00:00:19	7
27	ANLAGE EINSCHALTBEREIT	00:00:05	48
36	FREIGABE AUßENMASKE AUSGABE FEHLT	00:00:01	1

Abb. 9 Schichtauswertung Metallisierung

MBDE-SYSTEM V0.3   Tagesauswertung

Tagesauswertung   von : Mi   17.05.89   06:00:00   KW 20
                   bis : Do   17.05.89   06:00:00

ME-01-IL	Frühschicht	Spätschicht	Nachtschicht	Tag
Laufzeit	6:33:27	5:30:43	7:18:03	19:22:13
Stillstandszeit	1:26:33	2:29:17	0:41:57	4:37:47
Verfügbarkeit [%]	81.97	68.90	91.26	80.71
Stückzahl	8076	6759	8877	23712
Zykluszeit 1 [1] [s]	2.92	2.94	2.96	2.94
Zykluszeit 2 [2] [s]	3.57	4.26	3.24	3.64

LA-02-SE	Frühschicht	Spätschicht	Nachtschicht	Tag
Laufzeit	7:18:17	6:58:41	7:14:30	21:31:28
Stillstandszeit	0:41:43	1:01:19	0:45:30	2:28:32
Verfügbarkeit [%]	91.31	87.22	90.52	89.69
Stückzahl	13099	12383	11418	36900
Zykluszeit 1 [1] [s]	10.0	10.1	11.4	10.5
Zykluszeit 2 [2] [s]	11.0	11.6	12.6	11.7

SC-01-IL	Frühschicht	Spätschicht	Nachtschicht	Tag
Laufzeit	6:13:10	4:59:01	6:21:49	17:34:00
Stillstandszeit	1:46:50	3:00:59	1:38:11	6:26:00
Verfügbarkeit [%]	77.74	62.30	79.55	73.19
Stückzahl	14229	11340	13749	39318
Zykluszeit 1 [1] [s]	1.57	1.58	1.67	1.61
Zykluszeit 2 [2] [s]	2.02	2.54	2.09	2.20

[1] Zykluszeit 1 = Laufzeit / Stückzahl
[2] Zykluszeit 2 = Gesamtzeit / Stückzahl

Abb. 10 Tagesauswertung Metallisierung

Von der VAX werden die Schichtauswertungen und die Maschineneinstellungen der UMaKS ausgelesen und in eine Datenbank eingetragen. Dies geschieht vollautomatisch 1 mal am Tag durch einen Batch-Job. Aus den Datenbanken werden dann Schicht-, Tages-, und Wochenreports generiert.

In Abbildung 9 sieht man eine Schichtauswertung für die Metallisierung. Hier kann man genau sehen, warum die Sputteranlage stehen geblieben ist, und wieviel Zeit dies gekostet hat.

Die auf einen Tag verdichteten Daten sieht man in Abbildung 10. Dies wird man sich normalerweise am Terminal ansehen. Für die Anlagen, die besonders schlecht laufen, wird man sich dann eine Schichtstatistik erstellen lassen.

Die Daten kann man sich auch grafisch ausgeben lassen. Abbildung 11 zeigt die Stückzahlen der Kalenderwochen 46 - 51.

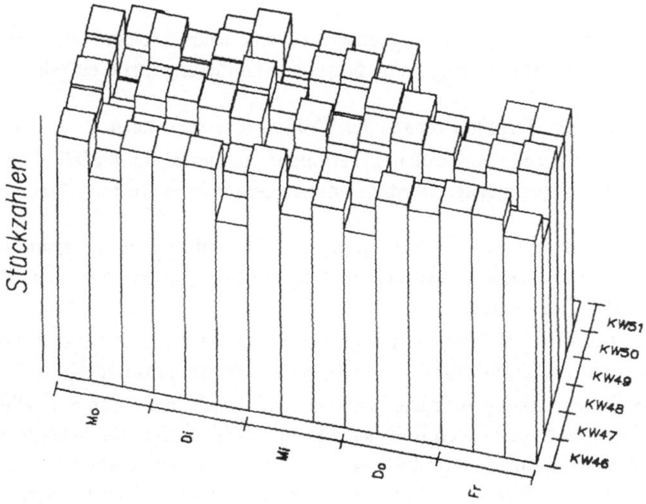

Abb. 11 Grafische Darstellung der Stückzahlen

Aufgetragen ist die pro Schicht produzierte Stückzahl an der Sputteranlage 1. Durch diese Darstellung erhält man schnell einen groben Überblick über die produzierten Platten.

```
SONOPRESS Maschinenzustand MBDE-SYSTEM    12. 6.89   14:29:01
```

	Zustand	Alarm	Zeit	Stückzahl	Zyklus [s]
SP09	läuft	–	14:29:00	227	9,68
SP14	steht	19	14:24:32	203	8,55
ME01	läuft	–	14:28:58	533	3,01
LA02	läuft	–	14:28:57	825	10,50
SC01	läuft	–	14:28:55	790	1,80
DR04	steht	17	14:27:01	651	2,00

Abb. 12 Anzeige Istzustand in der Pilotphase

Man kann sich den aktuellen Zustand der Maschinen auf dem Terminal zyklisch anzeigen lassen. Abbildung 12 zeigt einen Momentanzustand aus der Pilotphase. Hier waren zwei Spritzgußmaschinen, eine Metallisierung, eine Lackierung, ein Scanner und eine Druckmaschine angeschlossen.

In der Spalte Zustand kann man sehen, ob die Maschine momentan gerade steht oder läuft. Bei Alarm sind die gerade aktiven Alarme aufgeführt. Als Zeit ist die letzte von der Maschine beim UMaKS eingegangene Meldung angegeben. Unter Stückzahl findet man die in der aktuellen Schicht bearbeiteten Platten. Zyklus gibt die Zykluszeit der letzten bearbeiteten Platte an.

Wie die UMaKS sind die Terminals und Drucker am Terminalserver angeschlossen. Damit erreicht man eine große Hardware- und Software-Unabhängigkeit. Alle Möglichkeiten des MBDE-Systems stehen hardwaremäßig jedem User offen. Es ist eine reine Softwarefrage, wer auf welche Möglichkeiten zugreifen kann. Die Regulierung von Zugangsberechtigung, Verwaltung der Druckerqueues, benutzerabhängige Ausführung bestimmter Programme und benutzerabhängiger Schutz der Daten sind bereits im Betriebssystem VMS vorgesehen und brauchen nicht extra programmiert werden.

Um zu erreichen, daß das System möglichst auch von schon vorhandenen Rechnern mitbenutzt werden kann, wurden für alle Programme folgende Randbedingungen eingehalten :

- *Ausgaben auf dem Terminal verwenden VT100 Steuersequenzen.*
  Die meisten der bereits vorhandenen Terminals emulieren ein VT100. Die VT100-Terminalemulation ist auf den meisten Terminalemulatoren anderer Rechner vorhanden

- *alle Texte, die irgendwo gebraucht werden, Dokumentationen, Programme, Eingabedateien für EPROM-Programmiergeräte, Auswertungen sind entweder reine Text-Dateien ohne Steuerzeichen oder TEX Dateien.*
  Falls z. B. für EPROM-Programmiergeräte besondere Formate benötigt werden, gibt es ein Programm, das die Text-Dateien in das spezielle Format umwandelt.

  Durch die Einschränkung auf reine Text-Dateien erreicht man, daß man alle diese Informationen auf die unterschiedlichsten Rechner (mit unterschiedlichen Betriebssystemen, unterschiedlichen internen Darstellungen) übertragen (z. B. mit Kermit) und dort auch weiterverarbeiten kann (ausdrucken, editieren, einfügen in andere Programme). Dies bedeutet, daß praktisch alle Rechner, die über eine V.24-Schnittstelle verfügen oder einen Ethernetanschluß besitzen, angeschlossen werden können und die daran angeschlossene Peripherie mitbenutzt werden kann. Der Preis, den man dafür bezahlt, ist die relativ geringe Druck-Qualität der so produzierten Daten ("Teletype-Qualität").

  Um qualitativ hochwertige Auswertungen zu erstellen, wird TEX benutzt. TEX ist weit verbreitet, die Eingabe ist reiner Text und somit transportabel und TEX kann auch im Batch benutzt werden. Dies ist besonders für die automatisierten Auswertungen interessant. Die dabei für TEX notwendigen Steuerkommandos werden automatisch generiert.

- Terminalgrafik TEK4014 Format
  Für die Ausgabe von Grafiken auf den Bildschirm wird das Tektronix-4014-Format benutzt, das als ein "Quasi-Standard" von vielen Grafikterminals verwendet wird. Durch den Anschluß über eine V.24-Leitung kann solch ein Grafikterminal auch sehr weit entfernt aufgestellt werden.

- Grafiken werden im TR440 Vektorformat abgespeichert
  Die Gründe für dieses Format sind :
    - es ist einfach,
    - reiner Text (einfach zu transportieren),
    - Treiber für neue Geräte sind einfach und schnell zu implementieren,
    - dafür existieren Treiber zu anderen Protokollen und sehr vielen Geräten :
      TEKTRONIX 4014,HPGL, Kyocera Laserdrucker, Matrixdrucker
      AutoCAD, TEX - DVI-Treiber.

Zur Einrichtung eines MBDE Arbeitsplatzes stehen drei Stufen unterschiedlicher Funktionalität zur Verfügung :

- Textterminal oder Rechner mit Terminalemulator V.24
- Grafikterminal mit TEK 4014 Emulation und V.24 Anschluß
- Workstation über Ethernet

4. Besonderheiten in den einzelnen Fertigungsstufen :

Die bisher gezeigte Auswertung ist für alle Maschinen gleich. Der UMaKS erfüllt die Funktion des Schnittstellen-Wandlers und Datenverdichters. An einigen Beispielen soll jetzt demonstriert werden, daß das hier gewählte Konzept darüber hinaus auch auf die zum Teil sehr individuellen Besonderheiten der Maschinen eingehen kann.

Um auch auf Sonderwünsche schnell und flexibel reagieren zu können, wurde am UMaKS die Möglichkeit vorgesehen, Programme für Sonderauswertungen auf der VAX zu entwickeln und auf einen UMaKS zu transferieren. Diese werden dann in dem batteriegepufferten RAM des UMaKS gespeichert und sind auch nach einem Stromausfall dort noch vorhanden. Da der UMaKS 8 Tasks zur Verfügung stellt, ist für ein oder zwei Sonderprogamme bestimmt noch eine Task frei.

Grundsätzlich muß gesagt werden, daß in dieser CD Fertigung die Anlagen vom Maschinenbediener eingestellt werden. Dies unterscheidet diese CD Produktion von anderen Fertigungen, bei denen nur bestimmte - vorher festgelegte - Maschineneinstellungen von einem übergeordneten Rechner heruntergeladen werden können (CNC -Steuerung, Chip- Fertigung).

Am deutlichsten war diese unterschiedliche Strategie bei den Lackanlagen, die für den Einsatz in der Chip-Fertigung ausgelegt sind. Sie besaßen zwar alle Möglichkeiten, "remote" programmiert zu werden. Die Möglichkeiten, die Programme und Daten von einem übergeordneten Rechner zu holen, waren zwar im Protokoll vorgesehen, aber konkret nicht realisiert. Dies wurde vom Hersteller inzwischen nachprogrammiert.

4.1 Spritzguß

```
SONOPRESS                        Spritzguss KM 9                         MBDE-SYSTEM
00:36:05  25.03.89                  00:36  22:55  22:53  22:52  22:48  22:45  22:30
Schliesskraft          [kN]   xxxxx.........................................
Spritzdruck            [bar]  xxxxx.........................................
Nachdruck              [bar]  xxxxx.........................................
Staudruck              [bar]  xxxxx.........................................
Werkz-Sicherungsdruck  [bar]  xxxxx.........................................
Nachdruck  1           [bar]  xxxxx.........xxxxx...........................
Nachdruck  2           [bar]  xxxxx.........................................
Nachdruck  3           [bar]  xxxxx.........................................
Nachdruck  4           [bar]  xxxxx.........................................
Nachdruck  5           [bar]  xxxxx.........................................
Nachdruck  6           [bar]  xxxxx.........................................
Nachdruck  7           [bar]  xxxxx.........................................
Nachdruck  8           [bar]  xxxxx...........................xxxxx.........
Start schnell oeffnen  [mm]   xxxxx..xxxxx..................................
Start Nachdruck d. Weg [mm]   xxxxx.............................xxxxx..xxxxx
Plastifizierhub        [mm]   xxxxx...................xxxxx.................
Beg. Spritzg.-profil   [mm]   xxxxx...................xxxxx.................
```

Abb. 13 Einstellung an der Spritzgußmaschine

Aus Geheimhaltungsgründen werden die Tabelleneinträge nicht veröffentlicht. An der Spritzguß-maschine können vom Maschinenbediener über 200 Parameter frei eingestellt werden. Die Qualität der produzierten Platten hängt empfindlich von der Einstellung der Spritzgußmaschine ab. Es ist daher sinnvoll, die eingestellten Werte zu protokollieren, Einstelldaten bzw. deren Veränderung abzuspeichern. Bei Problemen kann dann auf diese Daten zurückgegriffen werden.

Ein systematischer Test zur Optimierung der Maschine ist überhaupt erst mit einem solchen automatischen System möglich, da sonst alle Parameter mitgeschrieben werden müßten. Dies ist sehr zeitaufwendig und fehleranfällig.

4.2 Metallisierung

Bei der Metallisierung muß das Target gewechselt werden. Die Target-Lebensdauer ist abhängig von der Anzahl der beschichteten Platten und der dabei verwendeten Sputterleistung. Die Target-Restlebensdauer wird von der SPS bei jedem Sputtervorgang automatisch erniedrigt und auf dem Display des Bedieners angezeigt. Da sich der Monitor im Reinstraum befindet, ist ein Ablesen immer mit Aufwand verbunden. Durch den Anschluß an einen UMaKS wird sie global verfügbar und auch grafisch auswertbar. Der im Computer gespeicherte Verlauf der Target-Restlebensdauer ermöglicht es, den Zeitpunkt für den nächsten Targetwechsel vorauszuplanen.

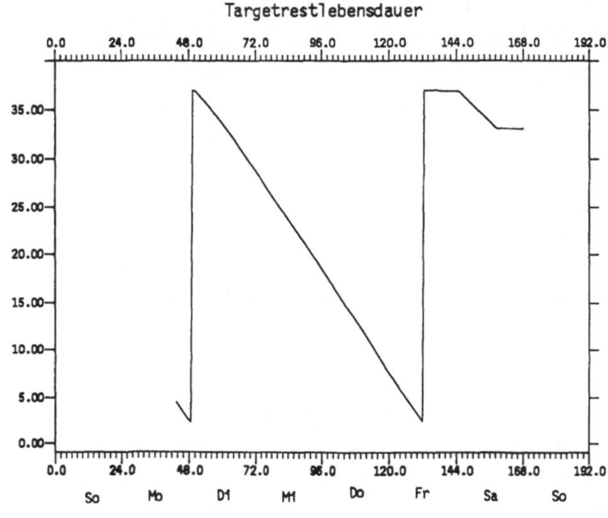

Abb. 14 Target-Restlebensdauer

4.3 Druck

Am Druck ist ein Klarschriftleser installiert, der sicherstellen soll, daß die Information auf der CD mit dem aufgedruckten Titel übereinstimmt. Auf jeder CD befindet sich eine Zahlenkombination, aus der der Inhalt eindeutig hervorgeht. Diese wird von dem Klarschriftleser ausgelesen und mit einem beim Auftragswechsel eingegebenen Sollstring verglichen. Stimmt dies nicht überein, wird die Maschine angehalten. Die Anzahl der überprüften Platten und die Häufigkeit von Fehlern werden schichtbezogen von einem Rechner abgespeichert. Als Klarschriftleser kommt ein komplettes OS-9 – System mit einer CCD-Kamera zum Einsatz. Daran angeschlossen ist eine Industrietastatur und ein Monitor, der im Normalfall das Bild der einzulesenden CD zeigt. Eine Anzeige von Statistiken war nur durch eine umständliche, blinde Tastatureingabe möglich. Dies erhöhte die Hemmschwelle zur Benutzung des System so sehr, daß die Statistiken fast nie ausgegeben werden. Durch den Anschluß des OS-9 Systems über einen zweiten V.24-Port an einen UMaKS sind auch diese Daten netzwerkweit verfügbar. Interessant sind hier die Schichtstatistik (gibt an, wieviele CDs in einer Schicht getestet wurden und wieviele davon fehlerhaft waren), und die Änderungsdatei, in der ein neuer Sollstring eingetragen wird. Bei der Eingabe eines neuen Sollstrings wird abgespeichert, wieviele Platten mit dem alten Sollstring getestet wurden und wieviele davon nicht übereinstimmen.

Im Beispiel in Abb. 15 kann man erkennen, daß der Sollstring falsch eingegeben wurde und daß dies zu 3 Fehlermeldungen des Klarschriftlesers geführt hat. Daraufhin hat der Maschinenbediener den Sollstring korrigiert.

Schichtauswertung Klarschriftleser :

	gesamt	gut	schlecht	error	
26.10.89 14:00	8803	8803	0	13	Frueh
26.10.89 22:00	8608	8575	33	2	Spaet
27.10.89 06:00	7718	7707	11	2	Nacht
27.10.89 14:00	9657	9656	1	8	Frueh

Änderungen des Sollstrings :

27.10.89 07:36	1047	1047	0	1	SONOPRESS C-9915 / CDV 2601 A
27.10.89 07:38	3	3	0	0	SCN)FFESS 260 240
27.10.89 08:18	994	994	0	2	SONOPRESS 260 240 A
27.10.89 09:04	657	657	0	0	SONOPRESS 259400 A

Abb. 15 Statistik Klarschriftleser

4.4 "Remote Measurement Call"

Da an den SPS - Steuerungen, die über den Paralleleingang am UMaKS angeschlossen sind, häufig Änderungen gemacht werden, gibt es auf dem UMaKS ein Programm, welches die Signalpegel am Parallelport mit einer maximalen Auflösung von 100 ms erfaßt. Diese werden dann im Speicher abgelegt und können grafisch ausgegeben werden (s. Abb. 16). Damit wird der UMaKS zu einem Protokolltester. Dabei ist zu betonen, daß die Datenerfassung während einer solchen Messung ungestört weiterläuft. Desweiteren existiert eine grafische Ausgabe für die A/D Wandlerkarte.

In Zusammenspiel zwischen der VAX und den UMaKS-Rechnern kann man von der VAX einen UMaKS anstoßen, eine Messung durchzuführen (SPS Timing-Diagramm oder A/D Wandler), nach der Messung die Daten zur VAX zu kopieren und diese Daten dann auf dem Bildschirm der Workstation darzustellen. In Analogie zum Aufruf von Unterprogrammen auf anderen Rechnern über ein Netzwerk ("remote procedure call") kann man dies als "remote measurment call" bezeichnen.

Das hier vorgestelle Konzept zur Realisierung einer Maschinendatenerfassung bildet ein in sich abgeschlossenes, eigenständiges System. Es besitzt zwei prinzipielle Schnittstellen, durch die es sehr leicht erweiterbar ist und durch die Interaktion mit anderen Systemen realisiert werden kann. Die eine Schnittstelle ist der UMaKS, durch dessen einfache Hardware- und Software- Erweiterbarkeit sich fast jede Maschine anschließen läßt.

Da man sich beim Nachrüsten der Maschinen mit entsprechenden Schnittstellen durch den UMaKS auf die jeweiligen Gegebenheiten des Maschinenherstellers einstellen kann (sowohl bei der Hardware- als auch bei der Software-Schnittstelle), spart man hier erhebliche Kosten.

Außerdem kann der Maschinenhersteller die gewünschten Änderungen konform zu seinem Protokoll vornehmen. Damit sind diese Änderungen auch in zukünftigen Updates des Maschinenherstellers enthalten, während sonstige Sonderlösungen bei einem Update der Maschinensteuerung zunächt einmal nicht berücksichtigt würden.

Durch den Einsatz einer VAX als übergeordneten Rechner eines Netzwerks auf Ethernetbasis, können die hier anfallenden Daten leicht anderen Rechnern zur Verfügung gestellt werden. Die Integration eines PC-Netzwerks und die Kommunikation mit UNIX Systemen sind somit möglich.

Durch den Anschluß eines Modems an die VAX ist ein Zugriff auf die Daten dieses System von außen möglich. Dies konnten wir auf der Ausstellung "Forschungsland Nordrhein-Westfalen" in Bonn (28.3.1990 - 4.4.1990) überzeugend demonstrieren.

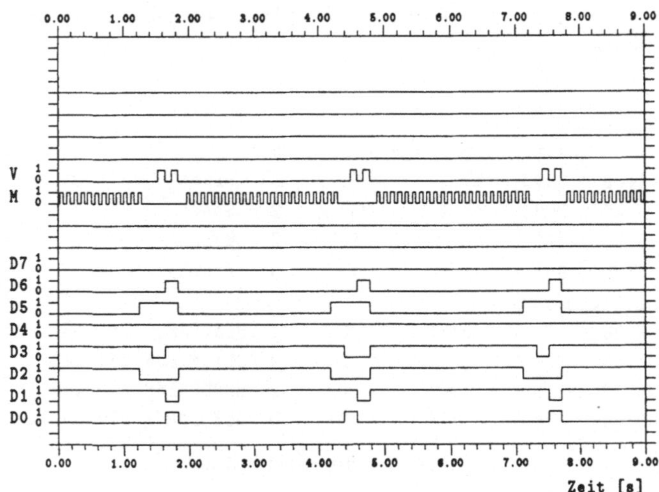

Abb. 16 SPS Timingdiagramm

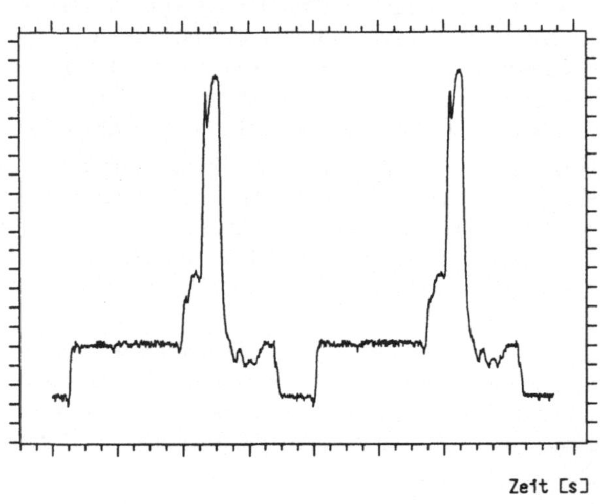

Abb. 17 Druckprofil der Spritzgußmaschinen

Die hier vorgestellte offene Architektur (offenes Betriebssystem, offenes Netzwerk, offene Maschinen-Schnittstellen, ...) ermöglicht eine einfache Integration anderer Netze, z. B. ließ sich das Netzwerk der Zentralen-Leittechnik von Landis & Gyr einfach in dieses System integrieren. Das ZLT - Netzwerk besteht aus einer zentralen PDP11, an die über mehrere Ringe Rechner ("Controller") angeschlossen sind, auf denen ein Multitasking-Basic-Betriebssystem läuft. Die an die "Controller" angeschlossen Geräte werden dann über mehrere Basic-Tasks gesteuert und überwacht.

Durch Anschluß eines Terminalausgangs an den Terminalserver ist jetzt hardwaremäßig ein netzwerkweiter Zugriff auf das ZLT-System möglich. Daten aus dem ZLT-System können jetzt mit

Kermit auf die VAX übertragen werden und dort weiterverarbeitet werden (z. B. grafisch aufbereitet auf dem Laserdrucker ausgegeben werden). Ein Beispiel hierfür ist die grafische Darstellung des Stromverbrauchs.

Der ZLT Anschluß ermöglicht das Austauschen der Basic-Programme und der auf der PDP11 erzeugten Daten. Das Bearbeiten, insbesondere der Basic-Programme, kann dann auf der VAX in einer wesentlich komfortableren Entwicklungsumgebung geschehen. Diese können dann auch auf dem Laserdrucker ausgedruckt werden.

Der Anschluß eines Modems an die VAX und das hier vorgestellte Netzwerk geben dem Maschinenhersteller die Möglichkeit, von draußen auf sein System zuzugreifen. Dies ist insbesondere zur Einstellung von Parametern, zur Einspielung neuer Software und zur Fehler-Ferndiagnose interessant.

Die Anforderungen an dieses System wurden speziell für die Bedürfnisse in der CD-Produktion entwickelt. Der in einer Pilotanlage realisierte Lösungsvorschlag ist aber so allgemein, daß er sich leicht auch auf andere Produktionen übertragen läßt. Diese Flexibilität ist notwendig, da sich die Anforderungen an ein solches System immer ändern werden. Nach der erfolgreichen Pilotinstallation wurde dieses System dann fabrikweit eingeführt.

5. Zusammenfassung

Dieses System macht durch eine kontinuierliche, automatische Datenerfassung eine objektive Beurteilung der Maschinen möglich. Es protokolliert alle an den Maschinen auftretenden Fehlermeldungen und ermöglicht deren zentrale Auswertung.

Durch eine Protokollierung der Maschineneinstellungen ist auch hinterher eine Fehleranalyse möglich. Durch die Blockfertigung können Fehler manchmal erst in späteren Fertigungsschritten erkannt werden. Dann laufen aber bei den Maschinen schon andere Aufträge.

Da alle Daten zentral gesammelt werden und für alle angeschlossenen Nutzer verfügbar sind, kann die Zahl der an den Maschinen installierten Protokolldrucker verringert werden. Die Reports werden automatisch erstellt. Dies spart Zeit (beim Generieren der Reports) und bringt die Information direkt zum Anwender (Terminal-Nutzer). Das Speichern der Daten über einen längeren Zeitraum und der Zugriff auf alle Daten macht maschinenübergreifende, automatisch generierte Reports möglich, die bisher wegen des erheblichen Aufwandes unterblieben sind. Die jetzt verwendeten Drucker stehen in Büroräumen und nicht mehr direkt neben den Produktionsmaschinen.

Durch das Netzwerk ist von jedem Terminal aus ein transparenter Zugriff auf die Maschinen - Einstellungen und -Parameter möglich. Dies ist ein wesentlicher Fortschritt, wenn man bedenkt, daß vorher diese Werte nur an den Terminals der Maschinen abrufbar waren. Diese befanden sich teilweise an schwer zugänglichen Stellen (Reinstraum) und sind teilweise durch Schlüsselschalter oder Paßworte gesichert. Waren bisher nur die momentanen Werte an den Maschinen vorhanden, so können jetzt die Maschinendaten eines größeren Zeitraums von vielen Terminals an unterschiedlichen Orten gleichzeitig kontrolliert und durch Programme analysiert werden.

Literatur:

[1] P. Diepstraten, G. Steemann, productronic 1/2 - 1988, 20
[2] Andrew, Tannenbaum, Operating Systems Design and Implementation
[3] W. Gerth, R. Hausdörfer, Zeitschrift für Computertechnik, 3 - 1985, 36
[4] W. Gerth, K.-H. Niemann, VMEbus 10 / 1989, 81
[5] J. E. McNamara, Technical Aspects of Data Communication, Digital Press, 1988

Ein Prozeduren-Satz zur asynchronen Ein-/Ausgabe in PEARL

Karlotto Mangold

ATM Computer GmbH

Bücklestraße 1-5

7750 Konstanz

Zusammenfassung:

Um Eingaben von mehreren asynchron sendenden Geräten in PEARL verarbeiten zu können, ohne die Nachteile der Durchführungsanweisungen bei Verwendung von READ, GET oder TAKE in Kauf nehmen zu müssen, wurde ein Satz von Prozeduren implementiert. Hier werden die Notwendigkeit für diese Implementierung, die Funktionalität der Prozeduren und Gesichtspunkte bei der Implementierung dargestellt. Zu diesen Implementierungsgesichtspunkten gehört auch die Darstellung der verwendeten Betriebssystem-Leistungen, die in PEARL nicht direkt als Sprachmittel vorhanden sind. Abschließend wird auf die Konsequenzen im Anwender-Programm im Gegensatz zum DATION-Betrieb hingewiesen.

Problemstellung

Betrachtet man die Anforderungen von Automatisierungs-Systemen an die verwendete Implementierungssprache und stellt diese Anforderungen den Möglichkeiten gegenüber, die PEARL bietet, so zeigt sich sehr schnell, daß die Parallelverarbeitung durch das Taskkonzept und die Taskoperationen von PEARL sehr gut abgedeckt wird. Auch der hardware-unabhängige Betrieb von Digital- und Analog-Baugruppen ist transparent und portabel möglich. Seit der Definition von PEARL hat sich jedoch die Struktur der zu betreibenden E/A-Geräte gewandelt. Heute sind viele Meßgeräte mit eigener Intelligenz ausgestattet und verfügen über Schnittstellen gemäß den CCITT-Normen V.24, V.28 und V.11. Damit könnten sowohl diese Geräte, als auch ein Großteil der Kommunikation zwischen Prozeßrechnern über die normalen Ein-/Ausgabeanweisungen PUT und GET in ähnlicher Weise betrieben werden wie ein Terminal. Der einzige Hinderungsgrund für einen solchen Betrieb ist die Tatsache, daß in PEARL alle Ein-/Ausgabeoperationen Durchführungsanweisungen sind, die erst verlassen werden, wenn die entsprechende Operation beendet ist. Während Ausgaben vom Rechner meist zeitlich überschaubar und planbar sind, erfolgen Eingaben häufig spontan und asynchron zur Ausführung der in den Verarbeitungsprogrammen vorgesehenen Eingabe-Anweisungen. Diese Asynchronität kann nun entweder dazu führen, daß eine Task beliebig lange in der Eingabeanweisung wartet oder daß die Eingabedaten verloren gehen, weil keine Eingabe Anweisung vorliegt, wenn die Daten anliegen. Da Ausgaben ohnehin meist über eine zentrale, beziehungsweise über eine gerätespezifische Ausgabetask (Spooler) abgewickelt werden, sollen im folgenden lediglich Eingabeaufträge betrachtet werden.
Häufig muß eine Vielzahl von Datenleitungen bedient werden, die unabhängig voneinander Meldungen liefern. Des-

halb führt der PEARL-gemäße Ansatz zu mindestens so
vielen Tasks, wie Leitungen bedient werden sollen. Oft
kommen nun von verschiedenen Leitungen dieselben Mel-
dungstypen, so daß bei meldungsspezifischer Verarbeitung
die entsprechenden Algorithmen für jede Leitung implementiert werden müssen, oder zusätzlich zur Eingabetask für
jeden Meldungstyp eine eigene Verarbeitungstask vorhanden sein muß. Dadurch wird die Anzahl der Tasks und
damit auch die Zahl der Taskwechsel weiter vergrößert.
Realzeitbetriebssysteme bieten heute oft die Möglichkeit,
asynchrone Ein-/Ausgabeaufträge zu formulieren, die lediglich zum Start eines Gerätes führen, den Auftraggeber-
Prozeß fortsetzen und nach Abschluß den Auftraggeber
informieren. Im Prinzip könnte diese Leistung in PEARL so
genutzt werden, daß mit der entsprechenden DATION ein IN-
TERRUPT verbunden ist, der ausgelöst wird, wenn der Auftrag beendet ist und der über eine WHEN-Einplanung zur
Aktivierung der entsprechenden Verarbeitungstask führt.
Diese Task könnte dann mit Hilfe einer GET-Anweisung die
Daten einlesen. Es ist jedoch nicht sichergestellt, daß
bei zeitlich dicht aufeinanderfolgenden Meldungen und
gleichzeitiger starker Rechner-Belastung durch das Anwen-
dersystem für jede Meldung ein Eingabeauftrag so recht-
zeitig anliegt, daß kein Datenverlust eintritt.
Zur Lösung dieses Problems wurde ein Prozedurpaket implementiert, das die Funktionalität des Betriebssystems für
den PEARL-Programmierer zugänglich macht.

Funktionalität der Prozeduren

Zusätzlich zu den üblichen Ein-/Ausgabeanweisungen wurden
Prozeduren definiert, die an den PEARL-DATIONs vorbei direkt mit dem Betriebssystem die entsprechenden Operationen abwickeln.
Neben einer Prozedur zur Initialisierung des Pakets wurden sechs Prozeduren definiert. Die Prozeduren OPN, CLS
und WRT sind funktional äquivalent zu den PEARL-Anweisun-

gen OPEN, CLOSE und PUT. Diese drei Prozeduren waren lediglich erforderlich, um den Betrieb in PEARL-ähnlicher Form zu ermöglichen, ohne die PEARL-Operationen mit den direkten Zugriffen auf dasselbe Gerät koordinieren zu müssen. Während die Prozedur WRT wie ein normales PUT wirkt und der Rücksprung zum Aufrufer erst erfolgt, wenn der Ausgabevorgang abgeschlossen ist, wurde die Eingabe in drei Aufrufe zerlegt. Selbstverständlich wäre die Implementierung der Ausgabe-Prozeduren in analoger Weise zu den nachfolgend beschriebenen Eingabe-Prozeduren möglich gewesen. Da aus der Anwendungssicht kein Bedarf dafür erkennbar war, unterblieb dies jedoch.

Die Prozedur STE (Starten der Eingabe) beschafft sich aus einem Puffer-Pool ein freies Pufferelement, in das die Daten eingelesen werden sollen, baut einen Eingabeauftrag auf und startet diesen. Dabei wird bei diesem Auftrag noch die Prozeß-Identifikation der Verarbeitungstask mitgegeben. Falls das Betriebssystem den Start von mehreren Aufträgen für dasselbe Gerät zuläßt, sind entsprechend viele STE-Aufrufe zulässig.

Die beim Eingabeauftrag bezeichnete Verarbeitungstask ruft die parameterlose Funktionsprozedur WIN (Warten auf Input) auf und wartet so auf den Abschluß eines Eingabe-Auftrages. Als Funktionswert liefert WIN eine Referenz auf das in STE beschaffte Pufferelement, das nun die Eingabedaten enthält. Dieses Pufferelement kann nun beliebig bearbeitet und an andere Tasks weitergegeben werden. Sind sämtliche Daten dieses Pufferelementes bearbeitet, so wird es durch den Aufruf der Prozedur FRE (Freigeben eines Elements) freigegeben und steht für die erneute Nutzung durch einen STE-Aufruf wieder zur Verfügung.

Bei diesem Eingabe-Konzept wurde ausgenutzt, daß das Betriebssystem für jeden beendeten Eingabeauftrag eine Rückmeldung an das Verarbeitungsprogramm sendet und daß diese Rückmeldungen im System solange gepuffert werden, bis sie mit einem Aufruf von WIN durch die bezeichnete Task abgeholt werden.

Implementierung der Prozeduren

Die sieben Prozeduren wurden vollständig in PEARL implementiert. Dabei erfolgt die Kommunikation mit dem Betriebssystem ATMOS [1] mit Hilfe des Intrinsics SVC, das es erlaubt, die standardmäßig vorhandenen Systemdienstaufrufe direkt von PEARL aus abzusetzen. Da die Prozeduren die Betriebssystem-Funktionen nutzen und dem Anwender PEARL-gemäß zur Verfügung stellen, sind sie, obwohl in PEARL programmiert, nicht ohne weiteres portabel. Die Programmierung erfolgte aus zwei Gründen in PEARL:

1. Durch die Existenz eines "PEARL-Beispiels" lassen sich auch durch den im Umgang mit dem Betriebssystem auf Assemblerebene ungeübten Anwender sehr einfach Modifikationen anbringen.

2. Die eigentliche Anwender-Software kann auf andere Systeme portiert werden, indem lediglich die Prozeduren entsprechend angepaßt werden.

Bei der Festlegung der Prozedur-Parameter wurde versucht, möglichst die entsprechenden PEARL-Typen zu verwenden. An zwei wesentlichen Punkten ergaben sich dabei Probleme:

1. Zur Identifikation der Verarbeitungstasks beim Aufruf von STE kann in PEARL der Task-Name nicht auf Parameterposition verwendet werden. Stattdessen wurde durch Aufruf einer herstellerspezifischen Standard-Funktionsprozedur dem Tasknamen die Prozeßnummer zugeordnet, unter der das Betriebssystem diese Task verwaltet. Diese Prozeßnummer wird als FIXED-Größe in PEARL definiert und als Parameter zur Task-Identifikation verwendet. Dabei kann der Anwender durch Aufruf dieser Standard-Prozedur für eine beliebige Task die Prozeß-Nummer ermitteln und diese als Verarbeitungstask anmelden.

2. Zur Identifikation des jeweiligen Gerätes wäre es im Prinzip möglich, den DATION-Bezeichner als Parameter beim Prozeduraufruf mitzugeben. Da jedoch bewußt darauf verzichtet wurde, diese Geräte bzw Leitungen sowohl über DATIONs als auch über diese Prozeduren zu betreiben, wurde zur Vermeidung von Verwechslungen auch hier auf DATION-Bezeichner verzichtet und an deren Stelle die logische Gerätenummer verwendet, mit der das Betriebssystem das jeweilige Gerät identifiziert und die dem Anwender aus seiner System-Konfigurierung ohnehin bekannt ist.

Der Vollständigkeit halber seien hier die Deklarationen der Prozeduren und eine kurze Beschreibung der Parameter angegeben:

OPN (in: logische Geräte-Nummer, Übertragungs-Geschwindigkeit, Parity, Datenbits/Zeichen, out: Status) zur Parametrierung und Eröffnung eines Anschlusses.

CLS (in: logische Geräte-Nummer, out: Status) zum Schließen eines Anschlusses.

WRT (in: logische Geräte-Nummer, auszugebende Datenstrecke vom Typ CHAR(), Anzahl der auszugebenden Zeichen, out: Status) zur Ausgabe eines Datenstrings über das adressierte Gerät.

WIN () ist eine parameterlose Funktionsprozedur, die als Funktionswert eine Referenz auf ein Pufferelement (siehe unten) liefert.

FRE (in: REF Puffer-Element, out: Status) zur Freigabe eines vom Anwender-Programm ausgewerteten Puffer-Elementes. Danach steht dieses Element zur erneuten Vergabe durch STE zur Verfügung.

STE (in: logische Geräte-Nummer, Prozeß-Nummer der Verarbeitungs-Task, out: Status) zum Beschaffen eines freien Puffer-Elements aus dem Puffer-Pool und Start eines Eingabe-Auftrags vom adressierten Gerät.

Neben diesen operativen Prozeduren ist der wesentliche Inhalt des Pakets der Puffer-Pool. Dies ist ein Array von Puffer-Elementen. Jedes Puffer-Element ist eine Struktur, die neben den Parametern des Eingabe-Auftrags und der Identifikation des zu startenden Prozesses den Speicherplatz für die einzulesenden Daten enthält. Diese Daten können mit den Sprachmitteln von PEARL so strukturiert werden, daß sie der Struktur der erwarteten Meldungen entsprechen. Damit können die eintreffenden Meldungen PEARL-gemäß verarbeitet werden.

In PEARL nicht vorhandene Leistungen

Im wesentlichen wurden bei der Implementierung dieser Prozeduren zwei Leistungen des Betriebssystems genutzt:

1. Starten eines Eingabeauftrags mit sofortiger Fortsetzung des aufrufenden Prozesses. Dabei bedeutet "sofortige Fortsetzung" natürlich lediglich den unmittelbaren Übergang in den Zustand "rechenwillig". Den Übergang in den Zustand "rechnend" und damit die Vergabe des Zentralprozessor steuert das Betriebssystem gemäß den Prioritäten aller rechenwilligen Prozesse.

2. Warten auf die Rückmeldung eines Eingabeauftrags. Diese Leistung könnte ebensogut durch die Aktivierung einer Task erzielt werden. Da jedoch das ACTIVATE in PEARL keine Möglichkeit der Informationsübergabe an die zu aktivierende Task kennt, mußte diese Leistung durch die Prozedur WIN zugänglich gemacht werden, da die Verarbeitungs-Task erfahren muß, welches Puffer-Element aktuell mit Eingabedaten gefüllt wurde.

Anwendung der Prozeduren

Geht man von sequentiellen Programmabläufen und den in PEARL als Durchführungsanweisungen vorhandenen Ein-/Ausgabeanweisungen aus, so erfordert die Anwendung der oben beschriebenen Prozeduren Änderungen im Design des Anwenderprogrammes. Durch die in PEARL üblichen parallelen Abläufe einerseits und die Asynchronität der Datenlieferanten andererseits sind jedoch keine grundlegenden Probleme bei der Verwendung der Prozeduren zu erwarten. Für die Anwendung kann im Prinzip folgende Programmstruktur empfohlen werden:

Zunächst werden die entsprechenden Anschlüsse durch Aufrufe von OPN parametriert und eröffnet. Dann wird für jeden Anschluß die entsprechende Anzahl von Eingabe-Aufträgen gestartet. Diese Anzahl läßt sich aus der Meldungshäufigkeit, beziehungsweise den Zeitabständen zwischen zwei Meldungen einerseits und der Zeit für die Meldungs-Bearbeitung andererseits so bestimmen, daß stets mindestens ein Eingabeauftrag gestartet ist.

Nach dieser Initialisierungsprozedur geht das Programm in eine Dauerschleife, die mit dem Aufruf von WIN beginnt. Wird bei der Eingabe kein Fehler zurückgemeldet, so erfolgt der erneute Start eines Eingabe-Auftrages mit dem Prozeduraufruf STE. Jetzt kann das bei WIN zurückgemeldete Pufferelement ausgewertet und bearbeitet werden. Sind alle Daten ausgewertet, so wird das Puffer-Element durch den Aufruf der Prozedur FRE freigegeben und steht für eine Wiederverwendung in einem späteren STE-Aufruf zur Verfügung. Damit ist auch der Schleifendurchlauf beendet.

Selbstverständlich kann auch die eigentliche Meldungsbearbeitung und die Puffer-Freigabe in einer anderen Task erfolgen, die nach einer kurzen Analyse der Meldung aktiviert wird. In diesem Fall muß bei der Taskaktivierung der Verweis auf das Puffer-Element in einer globalen Größe vom Typ REF Puffer-Element übergeben werden.

Literatur:

[1] ATMOS Betriebssystem für Computersysteme ATM 80/90, Konstanz, 1990

HPGL-Emulation in PEARL für Digitalplotter

J. Becker und W. Schepper

Fakultät für Physik
Universität Bielefeld
Universitätsstraße 25, D-4800 Bielefeld 1

Zusammenfassung

In diesem Artikel wird die Entwicklung eines Steuerungsrechners zur HPGL-Emulation auf einem Benson-Plotter beschrieben. Den Kern der Steuerung bildet ein Einplatinencomputer EPAC 68008 unter dem Betriebssystem RTOS-UH PEARL. Hauptpunkte dieses Artikels sind wichtige Programmteile wie Befehls-Scanner, Linien-Algorithmus, Clipping-Routine, Textzeichenausgabe mittels Bezierkurven und ein Einblick in die Programmentwicklung auf einem Entwicklungsrechner. Die Steuerung erreicht eine hohe Kompatibilität zu HPGL und nutzt die maximale Geschwindigkeit des Plotters von 2000 Schritten pro Sekunde voll aus. Durch die Entwicklung des Steuerungsprogramms in einer Hochsprache ist eine Programmwartung und -weiterentwicklung sowie die Anpassung an neue Aufgabenstellungen jederzeit zu erreichen. Denkbar wäre zum Beispiel auch der Einsatz der Steuerung als HPGL-Vorschaltgerät vor einem Laserdrucker, einem Bohrfräsplotter oder einer XYZ-Fräsanlage.

Einleitung

Es wurde unter Verwendung eines leistungsfähigen Einplatinenrechners mit einem modernen Prozessor ein Steuerungsrechner entwickelt und aufgebaut, auf dem Kommandos der Grafikbeschreibungssprache HPGL interpretiert und in die Inkrementalbewegungen eines Digitalplotters umgesetzt werden. Dabei konnte, unter Beachtung der speziellen Gegebenheiten des Plotters (Papierformat, Zeichengeschwindigkeit), eine möglichst hohe Kompatibilität zu HP-Plottern erreicht werden. In Abb. 1 findet man die bei uns eingesetzte Anordnung, bei der die von dem VAX-Rechnercluster kommenden HPGL-Plotfiles über den Steuerungsrechner in die XY-Bewegungen des Plotters (Benson Mod. 1212) umgewandelt werden. Das Terminal ist als Ersatz für das auf einigen HP-Plottern vorhandene LCD-Display anzusehen, auf dem Meldungen im Klartext ausgegeben werden können.

Softwareseitig stellen sich der Steuerung drei Aufgaben, die möglichst gleichzeitig bearbeitet werden müssen:

1. Kommunikation mit dem Rechner (Einlesen der Plotdaten),

2. Interpretation und Verarbeitung der HPGL-Befehle,

3. Ansteuerung des Plotters (Schrittmotoren und Stifte).

Durch die Verwendung des Multitasking-Betriebssystems RTOS-UH läßt sich die parallele Verarbeitung einfach durch Aufteilung in verschiedene Tasks erreichen. Hierbei wurden die Interpretation der Daten und die Ansteuerung des Plotters in einer Task zusammengefaßt, da bei einer Aufteilung die Verarbeitungsgeschwindigkeit durch zu häufige Task-Wechsel stark herabgesetzt wird. Wesentliche Punkte bei der Interpretation der HPGL-Kommandos sind neben der Zerlegung von Grafikprimitiven (Kreise, Kreisbögen, Rechtecke, usw.) und Textzeichen in Liniensegmente die damit verbundenen Probleme der Linienapproximation und des Clippings auf hardoder softwaremäßig vorgegebene Ausgabebegrenzungen. Der Artikel gibt einen Überblick über die **verwendete Hardware und interessante Aspekte der Software.**

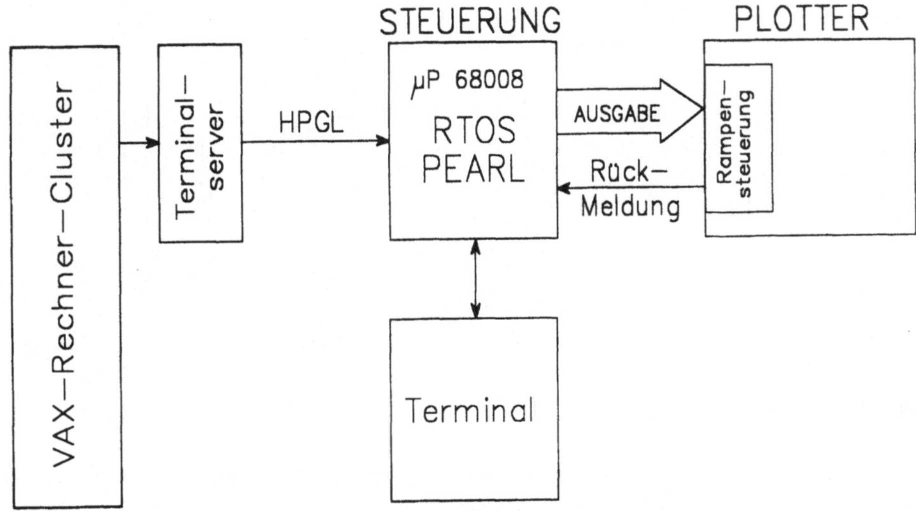

Abb. 1: *Anschluß der HPGL-Steuerung an Rechner und Plotter*

Der Steuerungsrechner

Auf Grund der Vorgaben (große Geschwindigkeit und hoher Kompatibilitätsgrad) wurde als Steuerungsrechner der in der Zeitschrift c't [6] vorgestellte Einplatinenrechner EPAC 68008 ausgewählt. Es gibt dafür mehrere Gründe:

1. die schnelle, moderne CPU MC68008,

2. die Anzahl der verfügbaren Schnittstellen und Ein- Ausgabeleitungen,

3. der ausreichend große Speicher.

Der wichtigste Punkt aber ist in der Implementation des Echtzeit-Multitasking-Betriebssystems RTOS-UH in Verbindung mit der Hochsprache PEARL auf dem EPAC 68008 zu sehen. Hiermit steht, mit Ausnahme eines Massenspeichers, eine Programmierumgebung zur Verfügung, wie man sie sonst nur auf einem Entwicklungsrechner findet. Das Steuerungsprogramm in einer Hochsprache erleichtert nicht nur die Programmentwicklung, sondern auch die Wartung und eventuelle Weiterentwicklung.

Der Steuerungsrechner ist zusammen mit dem Netzteil und zwei Interfaceplatinen in ein Halbschalen-Gehäuse eingebaut, auf dessen Frontseite sich neben Netzschalter, Netzkontrolleuchte und Sicherungshalter noch einige für den Betrieb wichtige Bedienungselemente befinden (siehe Abb. 2). Auf der Rückseite des Gehäuses sind die Anschlüsse für den Hostrechner, den Plotter und ein Terminal herausgeführt.

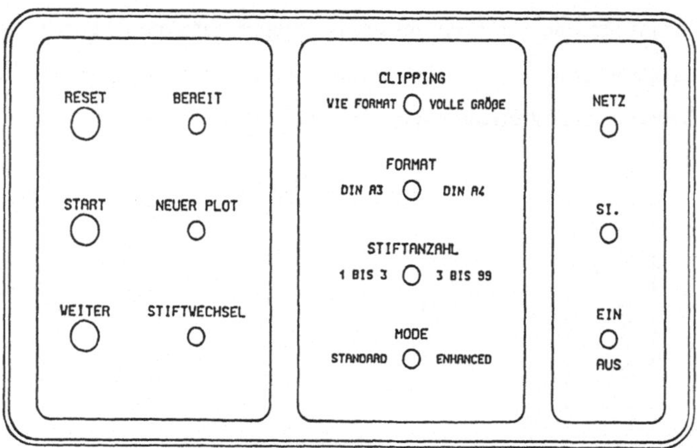

Abb. 2: Frontansicht der Steuerung mit Bedienelementen

Das Steuerungsprogramm

Der Schwerpunkt beim Aufbau der Steuerung liegt sicherlich in der Entwicklung der Software, denn hier wird die eigentliche Emulation des HPGL-Protokolls geleistet. Es ist in diesem Rahmen nicht möglich, auf alle Programmdetails einzugehen. Deshalb sollen hier nur der grobe Aufbau und die wichtigsten Routinen des Programms vorgestellt werden.

Der Systemteil des Programms

Im Systemteil werden die Interrupt-Routine, die seriellen Schnittstellen und die parallelen Ports konfiguriert. Die seriellen Schnittstellen verbinden die Steuerung mit dem Rechner und dem an der Steuerung angeschlossenen Terminal. Bei beiden wird der Datenstrom durch das XON-XOFF Protokoll gesteuert, bei der Terminalleitung wird auf Ctrl-A bis Ctrl-C, wie bei RTOS üblich, mit einer Eingabeanforderung geantwortet. Daten vom Rechner werden nicht interpretiert, sondern unverändert durchgereicht. Mit den I/O-Leitungen der parallelen Ports werden die Stellungen der Schalter an der Frontseite der Steuerung abgefragt (Eingabe), die Anzeigelämpchen bedient (Ausgabe) und die Befehle an den Plotter geschickt (Ausgabe). Die Interruptroutine wird zur Beendigung bestimmter Situationen benutzt, in denen die Steuerung auf eine Benutzerreaktion wartet (Starten eines Plots, Stiftwechsel, Fehlerzustände).

Struktur des Problemteils

Vor der detaillierteren Beschreibung des systemunabhängigen Problemteils wird zur besseren Übersicht einmal dessen grobe Struktur aufgezeigt.

Das eigentliche Programm besteht aus drei Tasks:

1. Die START-Task ist selbststartend und aktiviert nach der Initialisierung von wichtigen Variablen und Feldern die beiden anderen Tasks. Danach wird die Task durch ein TERMINATE beendet und erst nach einem RESET wieder aktiv.

2. Die FILLBUFFER-Task besteht nur aus einer Endlosschleife, in der Daten von der seriellen Schnittstelle empfangen und in einem Ringpuffer abgelegt werden.

3. In der SCANNER-Task befindet sich das eigentliche Hauptprogramm. Hier werden die Daten aus dem Ringpuffer entnommen, dekodiert, interpretiert und ausgeführt. In dieser Task finden sich auch die aufwendigen Programmteile, wie der Befehls-Scanner, die Clipping-Routine und die Textzeichen-Routine.

Funktion der START-Task

Hier wird nach einem Kalt- oder Warmstart die Steuerung in einen Grundzustand versetzt. So werden eine Einschaltmeldung an das Terminal abgesetzt, die Schalter an der Frontseite abgefragt und die Anzeigelämpchen gesetzt. Neben der Initialisierung von Feldern und Variablen und dem Einlesen von Fontdaten aus einem EPROM werden unter anderem auch eine Sinus- und eine Cosinustabelle angelegt. Diese Tabellen dienen speziell der Beschleunigung beim Zeichnen von Bögen und Kreisen, da hier viele Sinus- und Cosinusberechnungen durchgeführt werden müssen. Nach dieser Initialisierung werden die beiden anderen Tasks gestartet und die START-Task selbst terminiert. Durch die höhere Priorität der START-Task ist gewährleistet, daß die beiden anderen Tasks erst anlaufen, wenn die START-Task beendet ist und wertvollen Speicherplatz freigegeben hat. Während des normalen Programmlaufs wird diese Task nicht mehr aktiviert, sie wird nur beim Einschalten der Steuerung oder nach einer Betätigung des RESET-Tasters abgearbeitet.

Die FILLBUFFER-Task

In diesem Teil des Programms wird erstmalig die Multitasking-Fähigkeit von RTOS ausgenutzt. Um Unterschiede in den beiden Datenflüssen vom Rechner und zum Plotter auszugleichen, wird ein Ringpuffer (oder Umlaufpuffer) angelegt. Eine Task legt die ankommenden Daten im Puffer ab, während die andere sie dort zur Bearbeitung abholt. Im Idealfall sollten sich die beiden parallel laufenden Tasks dabei nicht gegenseitig behindern, daß heißt, die eine Task sollte immer dann laufen, wenn sich die andere gerade in einem Wartezustand befindet und umgekehrt. Hier kommt einem der Task-Scheduler von RTOS sehr entgegen, da er bei Tasks mit gleichen Prioritäten genau wie gewünscht verfährt. Da PEARL Semaphoren zur Tasksynchronisation zur Verfügung stellt, ist die Realisierung eines solchen Puffers sehr einfach und elegant und läßt sich in wenigen Zeilen formulieren:

```
FILL_BUFFER : TASK PRIO 20 RESIDENT;

  REPEAT
    REQUEST FREE;
    GET BUFFER(WRITE_POINTER) FROM INPUT BY A(1);
    WRITE_POINTER=(WRITE_POINTER REM BUF_LENGTH)+1;
    RELEASE FULL;
  END;
END;


SCANNER : TASK PRIO 20 RESIDENT;

  REPEAT
    REQUEST FULL;
    READ_CHAR=BUFFER(READ_POINTER);
    READ_POINTER=(READ_POINTER REM BUF_LENGTH)+1;
    RELEASE FREE;

        .
        .
        .

  END;
END;
```

Beide Tasks besitzen aus den oben genannten Gründen die gleiche Priorität und greifen über Schreib- und Lesezeiger auf den Puffer zu. Der Puffer hat eine Länge von (BUF_LENGTH) Zeichen. In der Schreib-Task wird ein einzelnes Zeichen von der Schnittstelle geholt, an die Position im Puffer geschrieben, auf die der Schreibzeiger weist, und danach der Schreibzeiger um 1 erhöht. Die Modulo-Berechnung (REM) des Zeigerwertes bewirkt, daß bei Überschreitung des Pufferendes der Schreibzeiger auf den Pufferanfang zurückgesetzt wird. Analog dazu wird in der Lese-Task verfahren, mit dem Unterschied, daß hier ein Zeichen aus dem Puffer gelesen wird.

Mit Hilfe der Semaphoren FREE und FULL wird verhindert, daß sich die Zeiger gegenseitig "überholen" und dabei entweder Zeichen verloren gehen, oder Zeichen gelesen werden sollen, die noch gar nicht im Puffer abgelegt wurden. Dazu wird die Semaphore FREE beim Programmstart mit dem Wert der Pufferlänge und FULL mit dem Wert Null vorbelegt. Die Summe der beiden Semaphorenwerte und damit die Summe aus freien und belegten Pufferplätzen kann den Pufferumfang nicht überschreiten und ein Über- oder Unterlauf des Puffers wird so wirkungsvoll verhindert.

Die SCANNER-Task

Mit der Beschreibung des Ringpuffers ist der erste Teil der SCANNER-Task behandelt. Auch weitere Programmteile sind Bestandteile dieser Task:

1. der Befehls-Scanner,

2. die Zeichenausgabe mittels Bezierkurven,

3. die Clipping-Routine,

4. der Linien-Algorithmus.

Aus Platzgründen wollen wir uns auf die Beschreibung der gerade genannten Routinen beschränken, wobei der Befehls-Scanner vollständig abgedruckt wird, weil er für eine korrekte und schnelle Interpretation der HPGL-Syntax sorgt und daher vielleicht von allgemeinem Interesse ist.

Der Befehls-Scanner

In diesem Programmteil wird der vom Hostrechner empfangene Datenstrom nach Befehlen und Parametern durchsucht, wobei die Parameter gleichzeitig von der ASCII-Textdarstellung in FLOAT-Zahlen konvertiert werden. Ist ein Befehl mit Parametern vollständig entschlüsselt, so wird er in der Prozedur CHOICE ausgeführt.

```
237    CODE=TOFIXED(READ_CHAR);
238    IF CODE>=32 THEN
239     IF CODE >= 97 AND CODE <=122 THEN  CODE=CODE-32;  FIN;
240     IF CODE >= 45 AND CODE <=57 THEN
241      IF NO_PARA THEN
242       PARACOUNT=PARACOUNT+1; VALUE=0.0;
243       SIGN=1; DECIMAL='0'B; NO_PARA='0'B;
244      FIN;
245      IF CODE >= 48 THEN
246       VALUE=VALUE*10+(CODE-48);
247       IF DECIMAL THEN SIGN=SIGN*0.1; FIN;
248      ELSE
249       IF CODE == 46 THEN DECIMAL='1'B;
250       ELSE
251        IF CODE == 45 THEN SIGN=-1; FIN;
252       FIN;
253      FIN;
254     ELSE
255      IF (CODE == 32 OR CODE == 44) AND NOT NO_PARA THEN
256       NO_PARA='1'B;
257       PARA(PARACOUNT)=VALUE*SIGN;
258       IF PARACOUNT == 100 THEN
```

```
259        CALL CHOICE(INST, PARA, PARACOUNT);
260        PARACOUNT=0;
261    FIN;
262    ELSE
263    IF CODE == 59 THEN
264    IF NOT NO_PARA THEN
265     NO_PARA='1'B;
266     PARA(PARACOUNT)=VALUE*SIGN;
267    FIN;
268    IF F == 3 THEN CALL CHOICE(INST, PARA, PARACOUNT); FIN; F=1;
269    ELSE
270    IF CODE >= 65 AND CODE <= 90 THEN
271    CASE F ALT INST.CHAR(1)=READ_CHAR; F=2;
272             ALT INST.CHAR(2)=READ_CHAR; F=3; PARACOUNT=0;
273                IF INST == 'LB' THEN
274                 CALL PRINT_LABEL; F=1;
275                ELSE
276                 IF INST == 'BL' THEN
277                  CALL BUFFER_LABEL; F=1;
278                 ELSE
279                  IF INST == 'DT' THEN
280                   CALL DEF_TERM; F=1;
281                  ELSE
282                   IF INST == 'SM' THEN
283                    CALL DEF_SYMB; F=1;
284                   ELSE
285                    IF INST == 'WD' THEN
286                     CALL PRINT_MESSAGE; F=1;
287                FIN; FIN; FIN; FIN; FIN;
288            OUT NO_PARA='1'B;
289                PARA(PARACOUNT)=VALUE*SIGN;
290                CALL CHOICE(INST, PARA, PARACOUNT);
291                INST.CHAR(1)=READ_CHAR; F=2;
292    FIN; FIN; FIN; FIN; FIN; FIN;
```

Das aus dem Ringpuffer gelesene ASCII-Zeichen wird zuerst in eine INTEGER-Zahl umgewandelt, welche in PEARL besser handhabbar ist. Durch die erste IF-Anweisung werden alle Zeichen mit Codes kleiner 32 unterdrückt, da diese keine Beachtung zu finden brauchen, und danach folgt eine Konvertierung von Klein- in Großbuchstaben. Die nun folgende IF–THEN–ELSE Verschachtelung wird hier durch die Reduktion auf die wichtigsten Teile etwas übersichtlicher dargestellt:

```
IF CODE >= 45 AND CODE <= 57 THEN
    ( Zeichen gehört zu einem Parameter: (+),(-),(.) oder(0-9) )
ELSE
  IF CODE == 32 OR CODE == 44 THEN ...
      ( Zeichen ist ein Separator ( ) oder (,) )
  ELSE
    IF CODE == 59 THEN
        ( Zeichen ist ein Terminator (;))
    ELSE
      IF CODE >= 65 AND CODE <= 90 THEN
            ( Zeichen ist ein Befehlszeichen (A-Z) )
  FIN; FIN; FIN; FIN;
```

Zeichen, die in keine dieser Kategorien fallen, bleiben unbeachtet. Nehmen wir einmal an, das Zeichen sei ein Befehlszeichen. Sollte es das erste der beiden Zeichen eines Befehls sein (F=1), so wird es in den String INST als erstes Zeichen eingetragen. Das zweite Zeichen (F=2) wird in INST an der zweiten Stelle eingetragen und dann die Parameteranzahl auf Null gesetzt. (Hier erfolgt dann noch eine Abfrage, ob es sich um einen der Befehle LB, BL, DT, SM oder WD handelt, da diese von dieser Stelle an anders behandelt werden müssen). Wird ein drittes Befehlszeichen empfangen (Terminator vergessen oder weggelassen), so wird dieses festgehalten, der Befehl als abgeschlossen betrachtet und nach der Berechnung des letzten Parameters erfolgt der Sprung nach CHOICE. Nach der Rückkehr aus CHOICE wird das festgehaltene Befehlszeichen als erstes Zeichen in INST eingetragen. Ist das Zeichen ein Terminator, so wird auch hier der letzte Parameter abgeschlossen und nach CHOICE verzweigt. Sollte das Zeichen ein Separator sein, der einem Parameter folgt, so wird der Parameter als abgeschlossen betrachtet. (Zusätzlich wird hier, falls eine Parameteranzahl von 100 erreicht wird, die Prozedur CHOICE angesprungen. Dies ist nötig, da im Feld PARA() nur für 100 Parameter Platz ist und nach der HPGL-Syntax unendlich viele Parameter zugelassen sind). Die letzte Möglichkeit besteht darin, daß das Zeichen zu einem Parameter gehört. Ist es das erste Zeichen eines Parameters, so wird die Parameteranzahl um 1 erhöht, der Wert des neuen Parameters auf 0 und der Vorzeichenfaktor auf 1 gesetzt und das DECIMAL-Flag gelöscht. Handelt es sich bei dem Zeichen um ein $(-)$, so bekommt der Vorzeichenfaktor den Wert -1 und bei einem Dezimalpunkt wird das DECIMAL-Flag gesetzt. Sollte das Zeichen eine Zahl (0 bis 9) sein, so wird diese zu dem, um eine Dezimalstelle nach links geschobenen, Wert von VALUE addiert und bei gesetztem DECIMAL-Flag zusätzlich der Vorzeichenfaktor korrigiert.

Implementation des Textausgabe-Befehls

Eine Hauptforderung an die Plottersteuerung war die Bereitstellung eines abgerundeten Zeichensatzes. Aus Kompatibilitätsgründen werden, nach anfänglichen Versuchen mit Splines und einem in [4] beschriebenen Algorithmus, nun Bezier-Kurven zur Abrundung des Zeichensatzes benutzt (Siehe auch Abb. 3 und Abb. 4). HPGL erlaubt mit dem CC-Befehl die Angabe des Abrundungsgrades der Zeichen (Abb. 5), und dies ist nur mit Hilfe der Bezier-Kurven zu erreichen. Dazu kommt, daß die Berechnung dieser Kurven recht einfach ist (siehe [5]) und daher auch sehr schnell erfolgen kann.

```
! " # $ %   ' ( ) * + , - . / 0 1 2 3 4 5 6 7 8 9 : ; < = >
? @ A B C D E F G H I J K L M N O P Q R S T U V W X Y Z [ \ ]
^ _ a b c d e f g h i j k l m n o p q r s t u v w x y z { | } ~
```

Abb. 3: Der vollständige ASCII-Zeichensatz in Originalgröße (A3)

Die Definition eines Textzeichens erfolgt durch die Festlegung von Punkten in einem Raster mit einer Höhe von 64 Punkten (+24 für Unterlängen) und einer Breite von 44 Punkten. Der Ausgangspunkt ist die linke untere Ecke des Rasters, die folgenden Punkte werden der Reihe nach

```
201 /*............................................................*/
202 /* FILLBUFFER(TASK): speichert Daten von der seriellen Schnittstelle */
203 /*                   im Ringbuffer.                                  */
204 /*............................................................*/
205 FILL_BUFFER : TASK PRIO 20 RESIDENT:
206   DCL WRITE_CHAR CHAR:
207
208   REPEAT
209     REQUEST FREE:
210     GET BUFFER(WRITE_POINTER) FROM INPUT BY A(1):
211     WRITE_POINTER=(WRITE_POINTER REM BUF_LENGTH)+1:
212     RELEASE FULL:
213   END:
214 END:
```

Abb. 4: *Beispiel für eine Textausgabe auf dem Plotter*

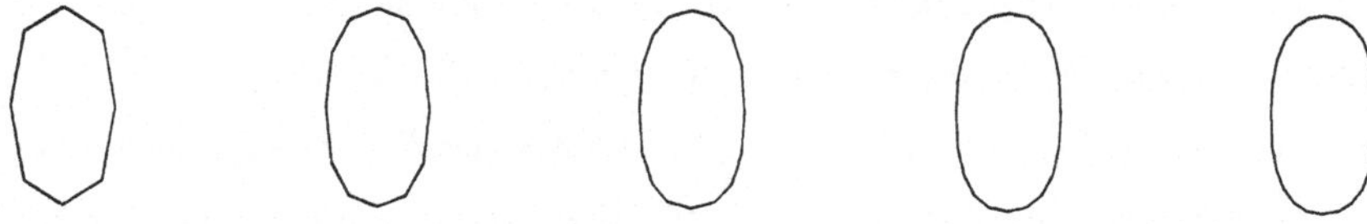

Abb. 5: *Verbesserung der Abrundung durch höhere Stützstellenanzahl*

mit gehobenem oder gesenktem Stift angefahren, wobei die Verbindung eine gerade Linie oder eine Bezier-Kurve (nur bei gesenktem Stift) sein kann.

Die Daten für die Textzeichen sind als Worte (16 Bit) hintereinander in einem EPROM abgelegt und aus Speicherplatzgründen recht sparsam kodiert. Die erste Zahl im Eprom ist die Angabe der Punktanzahl des Textzeichens. Dahinter folgen die Koordinaten der Punkte und die Art der Verbindung. Hierbei bestimmen die unteren sieben Bit die Y-Koordinate eines Punktes, die nächsten sieben Bit die X-Koordinate. Die Art der Verbindung ist in den beiden höchstwertigen Bits kodiert (siehe Tabelle 1).

Bit 15	Bit 14	Bedeutung
0	0	Linie mit gehobenem Stift
0	1	reserviert (3.Punkt)
1	0	Linie mit gesenktem Stift
1	1	Bezier-Kurve

Tabelle 1: *Kodierung der Verbindungsart in den höchstwertigen Bits*

Beim Programmstart oder auch bei einem Fontwechsel werden die Daten aus dem EPROM ausgelesen und in einem Array abgelegt. Hierbei werden gleichzeitig zwei Arrays mit Zeigern auf den Anfang und das Ende jedes Zeichens im Fontarray aufgebaut. Das Zeichnen der geraden Linien eines Buchstabens ist recht einfach. Nach der Dekodierung von Start- und Endpunkt der Linie wird nach Berechnung einer eventuellen Drehung oder Zeichenneigung die Linie durch den Aufruf der DO_LINE-Routine ausgeführt. Soll aber eine Bezier-Kurve gezeichnet werden (Bit 14 gesetzt), dann muß neben Anfangs- und Endpunkt der Linie noch ein dritter Punkt bestimmt werden. Um Speicherplatz zu sparen, ist dieser dritte Punkt nicht explizit angegeben, sondern wird auf eine recht einfache Weise bestimmt wie folgt: Stellt man sich Anfangs- und Endpunkt der Linie als diagonal gegenüberliegende Punkte eines Rechtecks vor, so bildet der, in Richtung der Linie gesehen, linke Eckpunkt des Rechtecks den gesuchten dritten Punkt. Die durch die drei Punkte bestimmte Bezierkurve führt also immer eine "Rechtskurve" aus. Die hierdurch entstehenden Einschränkungen bei der Konstruktion der Textzeichen sind aber minimal. (Auch diese Einschränkung läßt sich umgehen, indem unter Verwendung des reservierten Bitmusters der dritte Punkt explizit vorgegeben wird. Hiervon wird aber aus Speicherplatzgründen nur bei komplizierten Zeichen Gebrauch gemacht.) Nach der Berechnung einer eventuellen Zeichenneigung und -drehung werden dann in einer Schleife die Stützwerte der Bezierkurve bestimmt und die einzelnen Teilstücke der Kurve durch Aufruf von DO_LINE gezeichnet. Die Berechnung der Stützwerte geschieht recht schnell, da die Werte der Mischfunktionen an den Stützstellen schon vorberechnet sind.

Alles in allem zeigt sich, daß mit Hilfe von Bezier-Kurven eine sehr schnelle und effektive Ausgabe abgerundeter Textzeichen möglich ist. Darüber hinaus ist die Definition der Textzeichen recht einfach, und auch die Kompatibilität zu HPGL ist durch die Wahl einer variablen Anzahl von Stützstellen gewährleistet.

Die Clipping-Routine

Schon rein physikalisch ist eine Begrenzung des Plotbereichs durch das Papierformat und die Bewegungsgrenzen des Stiftes vorgegeben. Zusätzlich zu diesen Hardwaregrenzen kann man auch eigene Plotbegrenzungen durch die softwaremäßige Festlegung eines Ausgabefensters erreichen. Dies geschieht durch den Befehl **IW**(InputWindow). Alle Ausgaben, die nach der Festlegung eines Ausgabefensters folgen, werden auf diesen Fensterbereich begrenzt. Um das Verhalten des Plotters bei der Ausgabebegrenzung durch ein Clipping-Fenster zu beschreiben, muß man vier Typen von Linien unterscheiden (siehe auch Abb. 6):

1. Start- und Endpunkt liegen innerhalb des Fensters,
2. der Startpunkt liegt innerhalb, der Endpunkt außerhalb,
3. der Startpunkt liegt außerhalb und der Endpunkt innerhalb,
4. beide Punkte liegen außerhalb des Fensters.

Bei Linien vom Typ 1 bewegt sich der Stift genau wie ohne Clipping vom Start- zum Endpunkt. Linien vom Typ 2 werden bei gesenktem Stift bis zum Erreichen der Clipgrenze gezogen, dort wird der Stift gehoben und verbleibt so lange an dieser Stelle, bis ein weiterer Plotbefehl den Stift wieder in den Ausgabebereich bewegt. Bei Linien vom Typ 3 wird der Stift, unabhängig davon, wo er sich vorher befand, in gehobenem Zustand bis zum Schnittpunkt der Linie mit der Clipgrenze bewegt. Dort wird dann der Stift gesenkt und die Linie normal bis zum Endpunkt fortgeführt. Wie man in

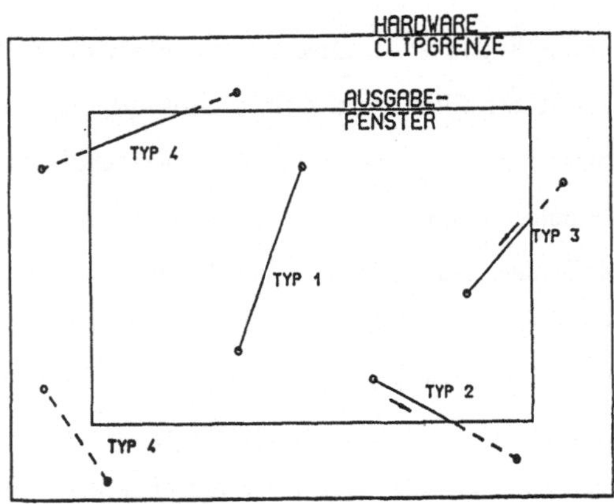

Abb. 6: Hard- und Softwareclipgrenzen

Abb. 6 sehen kann, gibt es für den Verlauf von Linien vom Typ 4 zwei Möglichkeiten. Existiert kein gemeinsamer Schnittpunkt zwischen Linie und Fenstergrenzen, so wird der Stift nicht bewegt. Im anderen Fall bewegt sich der gehobene Stift zum ersten Schnittpunkt, zieht von dort in gesenktem Zustand eine Linie zum zweiten Schnittpunkt, wird gehoben und verharrt dort.

Der im Programm verwendete Algorithmus ist im Prinzip aus dem Cohen-Sutherland-Algorithmus (beschrieben in [2]) entstanden. Jede Routine im Hauptprogramm, die eine Stiftpositionierung vornimmt, ruft diese Clipping-Routine mit den Parametern DX und DY auf. Hierbei handelt es sich um die Anzahl von Plotterschritten, die in die entsprechenden Richtungen ausgeführt werden sollen. Zuerst wird überprüft, ob beide Punkte innerhalb des Ausgabefensters liegen. Ist dies der Fall, so findet kein Clipping statt.

Muß die Linie geclippt werden, so wird nach dem Schema in Abb. 7 eine Bewertung der Linienendpunkte vorgenommen. Hierbei werden die X- und Y-Werte der Punkte einzeln bewertet. Sollte dabei festgestellt werden, daß beide X-Werte oder beide Y-Werte auf der gleichen Seite außerhalb des Fensters liegen, so liegt die ganze Linie außerhalb, und die Clip-Routine wird verlassen. Ist dies nicht der Fall, dann wird die Linie auf einen senkrechten Streifen eingeschränkt, indem die Schnittpunkte der Linie mit den Geraden berechnet werden, die das Clip-Fenster rechts und links begrenzen. Die Endpunkte der Linie werden durch die berechneten Schnittpunkte ersetzt.

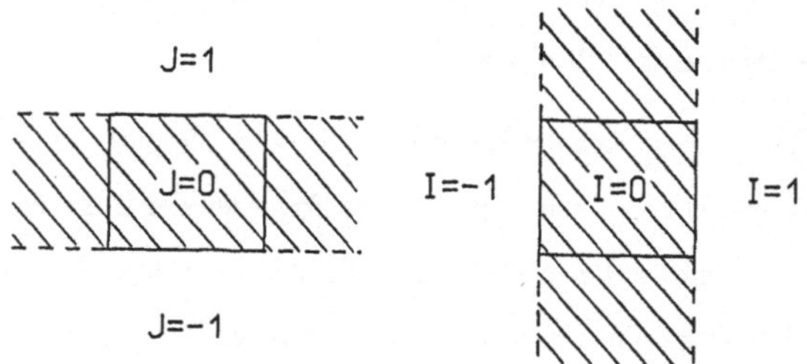

Abb. 7: Bewertung der Endpunkte getrennt für X- und Y-Werte

Liegt nach diesem ersten Clippen die Linie schon vollständig im Fenster, dann wird sie ausgegeben. Sollte dies noch nicht der Fall sein, dann wird nach einer erneuten Bewertung die Linie auch noch auf einen waagerechten Streifen eingeschränkt und die Schnittpunkte mit den beiden Geraden berechnet, die das Clip-Fenster nach oben und unten begrenzen. Die Endpunkte werden wieder durch die Schnittpunkte ersetzt, und nach einer erneuten Abfrage wird entschieden, ob Teile der Linie gezeichnet werden müssen oder nicht. Spätestens danach wird wieder zur aufrufenden Routine zurückgekehrt.

Der Linien-Algorithmus

Linien, die nicht mit einer der acht Basisrichtungen des Plotters zusammenfallen, müssen durch geeignete Aneinanderreihung von Schritten in diese acht Basisrichtungen approximiert werden. In Abb. 8 sieht man ein Beispiel für eine solche Linien-Approximation.

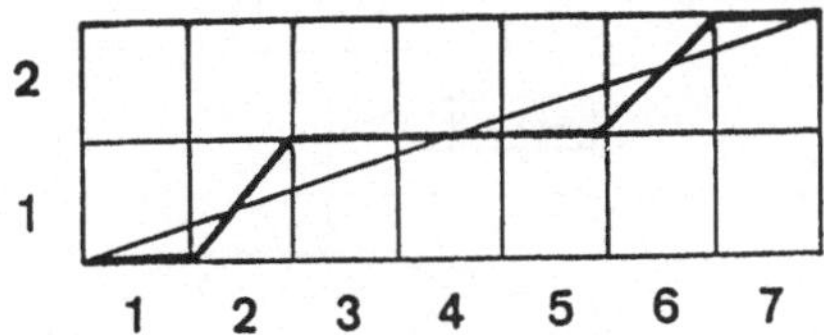

Abb. 8: Beispiel für eine Linienapproximation durch Basisschritte (aus [1])

Nach Versuchen mit eigenen Routinen wurde im Programm letztlich der in [1] vorgestellte Linien-Algorithmus verwendet. Da er in Form eines gut lesbaren Flußdiagramms vorliegt, wird dieses auf den folgenden beiden Seiten dargestellt.

Hierbei besitzen die Variablen folgende Bedeutung:

NX   Anzahl der Schritte in X-Richtung.

NY   Anzahl der Schritte in Y-Richtung.

NTOT Gesamtzahl der auszuführenden Schritte (ist gleich der größeren der beiden Zahlen NX und NY).

NDBL Anzahl der kombinierten Schritte (ist gleich der kleineren der beiden Zahlen NX und NY).

NSML Anzahl der einfachen Schritte. ($NSML = NTOT - NDBL$)

NS   Anzahl der hintereinander auszuführenden einfachen Schritte.

NL   Anzahl der hintereinander auszuführenden kombinierten Schritte.

NTES Testvariable, ob ein einfacher oder kombinierter Schritt erfolgen muß.

Der Linien-Algorithmus wird mit den Parametern X und Y aufgerufen, in denen die Anzahl der Schritte als FLOAT-Zahlen enthalten sind. Durch Rundung und Aufsummierung alter Restschritte (entstanden durch Rundung oder Clipping) wird die Anzahl auszuführender Voll-Schritte berechnet. Danach beginnt der Algorithmus, der im Flußdiagramm dargestellt ist. An den Stellen,

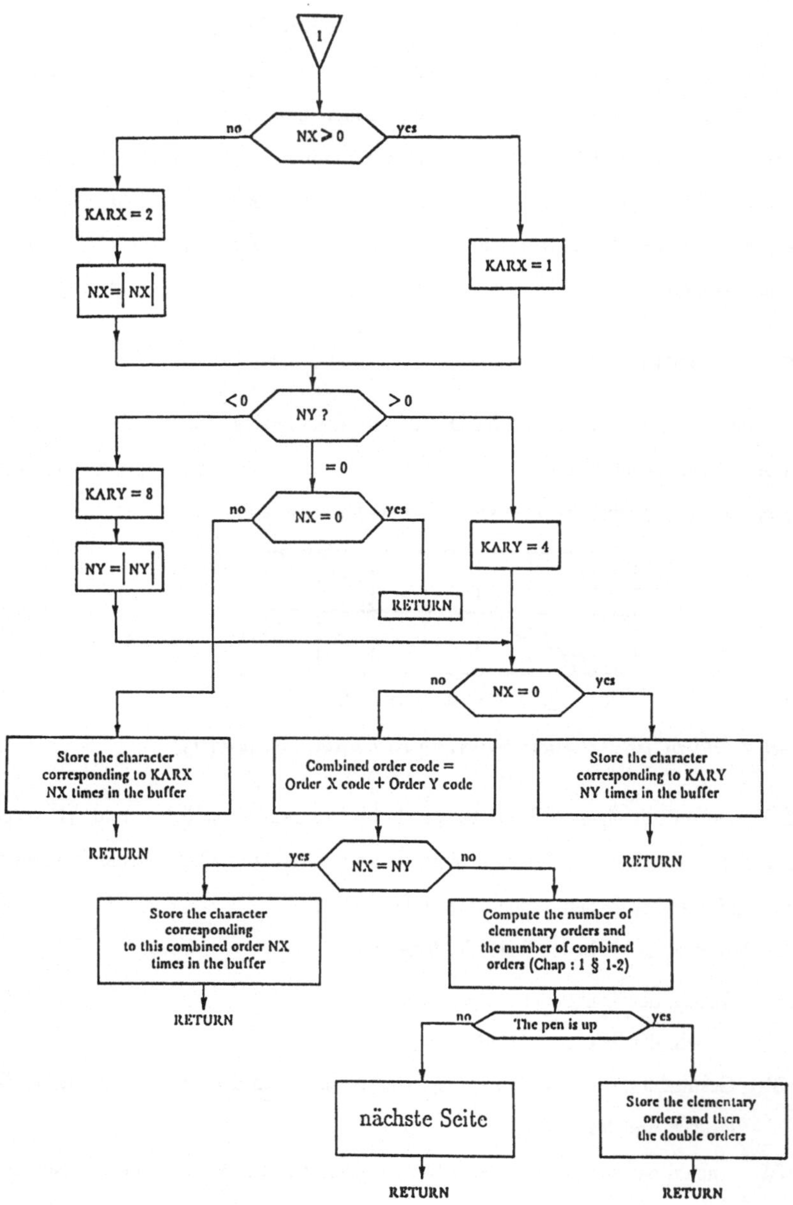

Abb. 9: Erster Teil des Linien-Algorithmus aus [1]

an denen im Flußdiagramm die Schritte in einen Puffer geschrieben werden, erfolgt im Programm ein Aufruf der WALK-Routine, in der entsprechend des jeweiligen Linientyps der Stift gehoben und gesenkt und die Schritte ausgeführt werden. Nach vollständiger Berechnung und Ausgabe der Linie wird noch eine Kennung für das Linienende zum Plotter geschickt.

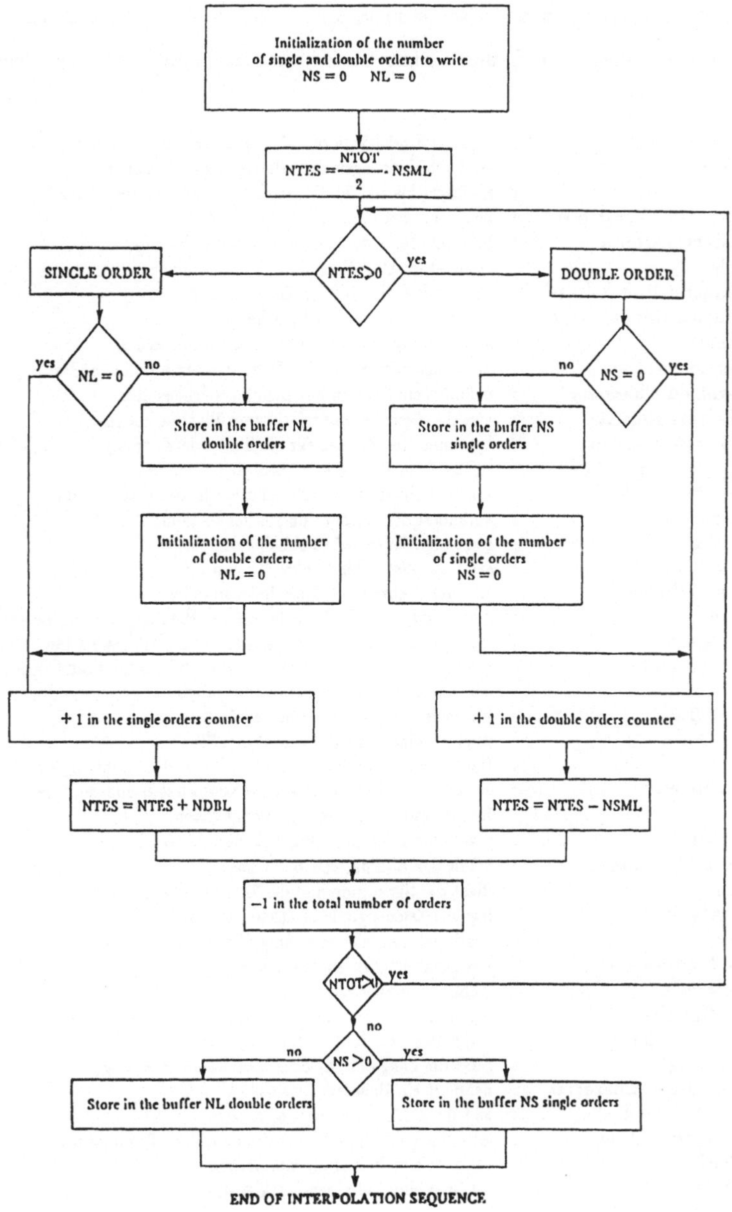

Abb. 10: Zweiter Teil des Linien-Algorithmus aus [1]

Liste der implementierten HPGL-Befehle in Kurzform

Hier sind in Kurzform die 40 Befehle aufgeführt, die von der Steuerung unterstützt werden. Die Befehle des HP7550, die von der Steuerung nicht ausgeführt werden können, sind zum größten Teil OUTPUT-Befehle, die nur bei einer interaktiven Benutzung des Plotters sinnvoll sind und mit dem Anschluß der Steuerung an eine Druckerqueue kollidieren. Einige weitere Befehle sind aus Speicherplatzgründen oder auch aufgrund der unterschiedlichen Hardware nicht zu implementieren.

AA Arc Absolute	:	zeichnet Kreisbögen mit absoluter Koordinatenangabe
AR Arc Relative	:	zeichnet Kreisbögen mit relativer Koordinatenangabe
BL Buffer Label	:	speichert bis zu 150 Textzeichen für eine spätere Ausgabe
CA Alternate Characterset	:	bestimmt den alternativen Zeichensatz
CC Character Chord	:	legt den Abrundungsgrad für Textzeichen fest
CI Circle	:	zeichnet einen Vollkreis
CP Character Plot	:	positioniert den Stift im Zeichenraster
CS Standard Characterset	:	bestimmt den Standard-Zeichensatz
DF Default	:	setzt wichtige Werte auf Standardwerte zurück
DI Direction Absolute	:	bestimmt den Winkel für Textausgaben absolut in Grad
DL Download Character	:	definiert ein Zeichen des Benutzerzeichensatzes
DR Direction Relative	:	legt die Textrichtung relativ zum Blattformat fest
DT Define Terminator	:	bestimmt das Zeichen für die Textende-Kennung
EA Edge Rectangle Abs.	:	zeichnet ein Rechteck in absoluten Koordinaten
ER Edge Rectangle Rel.	:	zeichnet ein Rechteck mit relativer Koordinatenangabe
ES Extra Space	:	verändert die Zeichen- und Zeilenabstände
IN Initialize	:	setzt alle Werte auf Einschaltwerte zurück
IP Input Points	:	setzt die Werte der Punkte P1 und P2
IW Input Window	:	legt die Grenzen des Ausgabefensters fest
LB Label	:	gibt Textzeichen in der eingestellten Richtung und Größe aus
LT Line Type	:	legt den Linientyp und die Länge des Linienmusters fest
PA Plot Absolute	:	zeichnet eine Linie in absoluten Koordinaten und setzt den Absolutmodus für Koordinatenangaben
PB Print Buffer	:	gibt einen vorher gespeicherten Text aus
PD Pen Down	:	zeichnet eine Linie mit gesenktem Stift
PG Page	:	führt einen Papiervorschub 1 cm hinter die Zeichnung aus
PR Plot Relative	:	zeichnet eine Linie in relativen Koordinaten und setzt den Relativ-Modus für Koordinatenangaben
PU Pen Up	:	positioniert den Stift in gehobenem Zustand
SA Select Alternate	:	wählt den alternativen Zeichensatz an
SC Scale	:	dient der Skalierung und der Definition von Benutzer-Einheiten
SI Size Absolute	:	legt die Zeichengröße in Zentimetern fest
SL Slant	:	bestimmt die Zeichenneigung (für kursive Schrift)
SM Symbol Mode	:	bestimmt ein Zeichen für den Symbol-Modus
SP Select Pen	:	wählt einen Stift an
SR Size Relative	:	bestimmt die Zeichengröße relativ zum Blattformat
SS Select Standard	:	wählt den Standard-Zeichensatz an
TL Tick Length	:	setzt die Länge für X- oder Y-Achsenmarkierungen
UC Userdefined Character	:	definiert ein Benutzer-Zeichen
VS Velocity Select	:	legt die Zeichengeschwindigkeit fest
WD Write to Display	:	gibt einen Text (bis 80 Zeichen) auf dem Terminal aus
XT X-Tick	:	zeichnet eine X-Achsen-Markierung
YT Y-Tick	:	zeichnet eine Y-Achsen-Markierung

Programmentwicklung auf dem EPAC

Die Entwicklung von Programmen für Einplatinenrechner, bei denen weder Benutzerinterface noch Massenspeicher zur Verfügung stehen, erfolgt im allgemeinen in Assemblersprache auf einem Entwicklungsrechner. Wenn in dem Programm Hardwareadressen oder Ein/Ausgabebausteine angesprochen werden, die auf dem Entwicklungsrechner nicht vorhanden sind oder zumindest an anderer Stelle liegen, wird die Programmentwicklung erheblich erschwert. Die Entwicklung komplexer Programme ist so nicht mit einem vertretbaren Zeitaufwand möglich.

Diese Probleme sind beim EPAC 68008 durch die Implementation von RTOS-UH/PEARL im wesentlichen beseitigt. Für die Programmentwicklung gibt es mehrere Möglichkeiten, da RTOS den Anschluß von Rechnern oder Terminals an beiden seriellen Schnittstellen gestattet. Schließt man an eine der seriellen Schnittstellen ein Terminal an, so können kleinere PEARL- oder Assemblerprogramme mit Hilfe des eingebauten Editors direkt auf dem EPAC geschrieben werden. Mit dem, ebenfalls im EPROM befindlichen, PEARL-Compiler oder Assembler wird das Programm übersetzt, und die vom Compiler erzeugten S-Records werden vom Linker/Lader in ablauffähiger Form im Speicher abgelegt. Sollten beim Test des Programms Fehler auftreten, so kann nach einer Änderung sofort neu compiliert, geladen und wieder getestet werden. Die Länge eines derartig erzeugten Programms ist durch den begrenzten Speicherplatz auf dem EPAC (ca. 20 kByte freies RAM) allerdings auf wenige kByte limitiert, da das Programm im Prinzip in drei verschiedenen Formen Speicherplatz belegt (Programm, S-Record-Datei und ablauffähiger Code). Außerdem geht das Programm nach dem Abschalten der Spannungsversorgung des EPAC verloren, es sei denn, man sendet es über die zweite serielle Schnittstelle zu einem Rechner mit Massenspeicher und sichert es dann dort.

Da mit dem ATARI ST unter RTOS-UH/PEARL ein Entwicklungsrechner mit erheblich mehr Speicherplatz zur Verfügung steht, kann man aber auch ganz anders vorgehen. Die erste Möglichkeit ist, den Programmtext auf dem ATARI zu schreiben und ihn dem PEARL-Compiler auf dem EPAC über die serielle Schnittstelle anzubieten. Auf diese Art belegt dann wenigstens der Programmtext im EPAC keinen wertvollen Speicherplatz mehr. Man kann aber auch gleich das Programm mit dem PEARL-Compiler des ATARIs übersetzen und nun die S-Records dem Lader/Linker über die serielle Schnittstelle anbieten. Zusätzlich kann in diesem Fall das Compiler-EPROM auf dem EPAC durch einen RAM-Baustein ersetzt werden, und man erhält so noch einmal 32 kByte freien Speicher. Diese letzten beiden Möglichkeiten wurden während der ersten Hälfte der Entwicklung des HPGL-Emulators genutzt.

Bis hierher geht aber das Programm noch bei jedem Abschalten der Spannungsversorgung verloren. Für den späteren Einsatz der Steuerung ist es natürlich notwendig, daß das Programm sofort nach dem Einschalten der Stromversorgung zur Verfügung steht und auch von selber anläuft. Hier bietet RTOS mit dem PROM-Befehl die Möglichkeit, das Programm in eine direkt im EPROM ablauffähige Form umzuwandeln. Dabei werden Programm- und Datenbereiche entsprechend den im Programmkopf anzugebenden Adressen von RAM- und EPROM-Bereich getrennt und der Code angepaßt. Mit dem AUTOSTART-Befehl kann der Code zusätzlich so modifiziert werden, daß das Programm gleich nach dem Systemstart losläuft. Von den beiden, vom PROM-Befehl erzeugten, S-Record-Blöcken wird einer auf die im Kopf angegebene und der andere auf eine beliebige Adresse ins EPROM gebrannt und dieses EPROM in den EPAC gesteckt. Hier bietet sich entweder der Steckplatz des nicht mehr benötigten PEARL-Compilers an oder man schafft Platz im Betriebssystem-EPROM, indem man nicht benötigte Module entfernt. Wir haben das letzte Verfahren gewählt, da das EPROM, welches den Compiler ersetzt, die Daten für verschiedene Fonts aufnehmen sollte. Im Betriebssystem-EPROM verbleibt nur der geschützte Bereich mit NUKLEUS, ERROR, IMP, SHELL und HYP und zusätzlich das MATH-Modul. Damit stehen ungefähr 29 kByte für das Emulationsprogramm zur Verfügung.

Die letzte Entwicklungsphase des Programms fand vollständig auf dem ATARI statt. Hier taucht nun wieder das am Anfang des Kapitels beschriebene Problem der unterschiedlichen Adresslage auf Ziel- und Entwicklungsrechner auf. Bevor man das Programm mit dem PROM-Befehl bearbeiten und in das Eprom brennen kann, müssen ja erst mit dem Linker/Lader die Verbindungen zu den hardwareabhängigen Systemadressen hergestellt werden, die auf beiden Rechnern nicht dieselben sind. Dieses Problem umgeht man, indem man dem Lader die Systemadressen des EPAC aufzwingt. Hierzu erzeugt man sich eine Laderliste mit Systemadressen des EPAC (siehe hierzu [7]).

Diese Datei wird nun mit dem Hauptprogramm gelinkt, und danach wird - wie weiter oben beschrieben - verfahren. Um sich von der Vorgehensweise eine bessere Vorstellung machen zu können, wird hier einmal ein kompletter Zyklus einer Programmänderung vorgestellt:

1. Änderung des Programms mit **ED** *myprog*. Durch das Verlassen des Editors wird das geänderte Programm automatisch gesichert.

2. Übersetzen mit (z. B.) **P.X LO=NO CO=OUTPUT** *myprog*. Ruft den PEARL-Compiler auf und erzeugt einen Prozeß mit dem Namen **X**. Es wird kein Listing-Output erzeugt. Die erzeugte S-Record-Datei bekommt den Namen OUTPUT.

3. Laden und Linken mit **LOAD OUTPUT+CROSS**. Hierbei ist CROSS die Laderliste mit den Adressen des EPAC 68008.

4. Umwandlung in ein selbststartendes Programm mit **AUTOSTART** *myprog*.

5. Erzeugen ROM-fähigen Codes durch **O ROM--PROM** *myprog*. Der Output von PROM wird in die Datei ROM umgeleitet und die durch den PROM-Befehl erzeugten zwei S-Record-Blöcke in dieser Datei abgelegt.

6. Da der uns zur Verfügung stehende Eprom-Brenner Binärdaten erwartet, werden mit einem Hilfsprogramm die beiden S-Record-Blöcke getrennt und in Binärdaten gewandelt. Zusätzlich müssen die Dateien vom RTOS-Diskettenformat über das ATARI-Format in das IBM-Format übertragen werden, da der Eprommer an einem IBM-PC betrieben wird.

7. Als letzter Punkt erfolgt das Brennen der beiden Datenblöcke in ein EPROM, welches dann in den EPAC gesteckt werden kann.

Der Durchlauf eines solchen Zyklus dauert etwa 30 bis 40 Minuten, wobei die Format-Konvertierung und das Brennen des EPROMs die meiste Zeit in Anspruch nehmen.

Literaturliste :

1. Benson Handbuch, *Basic Programs.*
2. Harrington, S., *Computergrafik — Einführung durch Programmierung*, MacGraw Hill, Hamburg, (1988).
3. Hewlett-Packard Handbuch, *Interfacing and Programming Manual HP 7550A.*
4. Jordon, B. W., Lennon, W. J., Holm, B. D., *An Improved Algorithm for the Generation of Non-parametric Curves*, IEEE-Transactions on Computers, Vol. C-22, No. 12, (1973).
5. Newman, W., und Sproull, R., *Grundzüge der interaktiven Computergrafik*, MacGraw Hill, Hamburg, (1986).
6. Schäfer, B., und Assenbaum, J., *Zwergenaufstand—EPAC 68008-klein, aber oho!*, c't, Heft 2 1987, Verlag Heinz Heise, Hannover.
7. Weitz, C. M., *Echtzeit-Multitasking mit RTOS/PEARL—PEARL-Programme in EPROMs*, c't Heft 2 1987, Verlag Heinz Heise, Hannover.

Ein intelligentes und netzfähiges BDE-Terminal
auf der Basis eines Motorola-Rechners mit dem RTOS-Betriebsystem.

Manfred Gerlich
Gesellschaft für Mikroelektronik und Systemtechnik mbH
Jurastr.16, 7000 Stuttgart 80

Andreas Nowakowski
Institut für Regelungstechnik und Prozeßautomatisierung
Universität Stuttgart Pfaffenwaldring 47, 7000 Stuttgart 80

Zusammenfassung

Auf der Basis eines 68008 Motorola Prozessors wurde ein Betriebsdatenerfassung-Terminal für den industriellen Einsatz entwickelt. Die Entwicklung mußte eine Reihe von Anforderungen berücksichtigen, die einerseits durch die vorgesehenen Anwendungsgebiete und Funktionen, andererseits aus Termin- und Kostengründen entstanden sind. Zu den wichtigsten hardwaremäßigen Eigenschaften des Terminals zählen seine Vernetzungsfähigkeit, Anschlußmöglichkeit verschiedener peripherer Geräte und der direkte Anschluß von Prozeßsignalen. Die Software ist echtzeitfähig und garantiert kurze Reaktionszeiten. Bei der Softwareerstellung wurde das RTOS-Betriebssystem eingesetzt und um notwendige systemspezifische Treiber erweitert, die Programmierung erfolgte grundsätzlich in der PEARL-Programmiersprache (mit der Ausnahme der hardwarenahen Systemsoftware). Nach kurzem Überblick des Anwendungsgebietes beschäftigt sich der Vortrag mit den zu berücksichtigen Anforderungen und stellt die Hardware- und Systemsoftware-Lösung dar. Das angewandte Kommunikationsprinzip wird genau beschrieben. Es werden auch die aus dem praktischen Einsatz entstandenen Erfahrungen und kurze Bewertung der Lösungen beigefügt.

1. Zielsetzung

Hierarchische rechnergestützte Lager- und Transportsysteme werden zweckmäßigerweise mit mehreren Steuerungsebenen realisiert. Das Bild 1 zeigt die Ebenenaufteilung und grobe Funktionen, die durch diese Ebenen zu erfüllen sind.

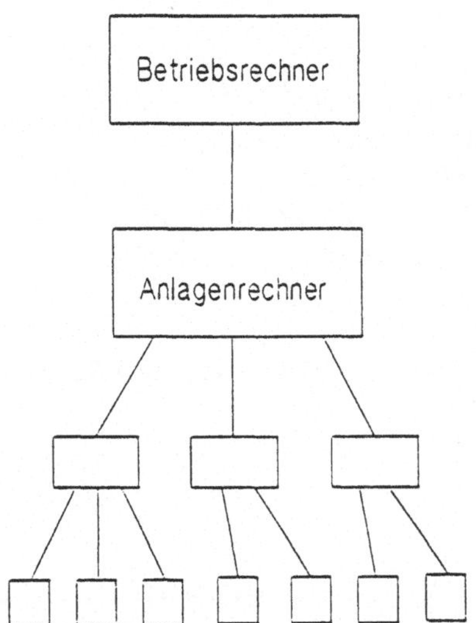

Bild 1 - Steuerungsebenen und grobe Funktionen

Diese Ebenen werden mit unterschiedlichen Systemen realisiert :
- die unterste Ebene wird als kontaktbehaftete Steuerung ausgeführt,
- die Ebene der Ablaufsteuerung wird oft als SPS (z.B. Simatic S5) ausgeführt,
- auf der Anlageebene - softwaremäßig mit Hilfe eines Anlagenrechners mit einem echtzeitfähigen Betriebssystem (Prozeßrechner PDP-11/RSX, VAX/VMS),
- auf der Betriebsebene - softwaremäßig mit Hilfe eines Betriebsrechners (VAX/VMS, IBM, Tandem - oft Mehrrechnersysteme).

In automatischen Lagersystemen wird das Material oft an besonders gestalten Übergabeplätzen an der Regallängsseite bereitgestellt. Neben den für die sicherheitstechnischen Belange dieser Plätze zuständigen Elementen ist auch, bedingt durch die Vielzahl dieser Plätze, eine umfangreiche Peripherie zu installieren und zu steuern. Dies führt bei zentralisierten Systemen zu sehr großen Steuerzentralen. Neben diesen Steuerungsaufgaben muß auch der synchron zur Anlage ablaufende Informationsfluß in das Gesamtkonzept integriert werden. Aus den genannten Gründen entstand die Notwendigkeit spezielle BDE-Terminals für dezentrale Datenerfassung und lokale Steuerung für die Übergabeplätze anzuwenden. Durch die für ein Projekt durchgeführte Marktanalyse haben wir leider feststellen müssen, daß die damals verfügbaren Geräte unseren Bedürfnissen nicht entsprachen. Deshalb haben wir uns entschlossen, ein eigenes BDE-Terminal zu entwickeln.

Das BDE-Terminal soll für die lokale Steuerung technischer Teilprozesse eingesetzt werden, wobei der Prozeßzustand lokal bearbeitet und gespeichert wird. Alle durch das BDE-Terminal vom Menschen eingegebenen materialbezogenen und auftragsbezogenen Daten (z.B. Materialanforderung, Identifizierung, Mengenangaben), werden zur nächsten Steuerebene (Anlagenrechner) weitergeleitet.

Der Anlagenrechner :

- empfängt Prozeßzustände von BDE-Terminals,
- empfängt Daten anderen dezentralen Steuerungen (z.B. SPS),
- schleust die material-/auftragsbezogene Daten zur Betriebsebene durch,
- empfängt die Ein-und Auslageraufträge von der Betriebsebene,
- sendet Daten und zu realisierende Steuerungs-Funktionen an die BDE-Terminals,
- überwacht Aktivitäten der anderen dezentralen Steuerungen,
- beauftragt die Regalbediengeräte.

Das Bild 2 stellt die Informationsflüsse zwischen Menschen, Teilprozessen, BDE-Terminals und dem Steuerrechner dar.

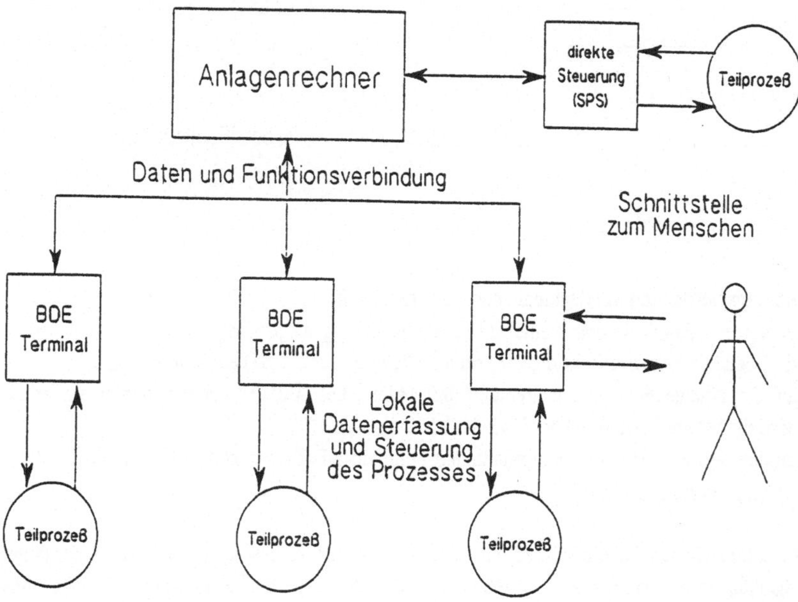

Bild 2 - BDE-Terminals, Anlagenrechner und dezentraler technischer Prozeß

2. Anforderungen an die BDE-Terminals und ihre Entwicklung

Die Entwicklung der BDE-Terminals mußte folgende Anforderungen berücksichtigen :

Allgemeines :

- Hard- und Software sollen möglichst parallel entwickelt werden können,
- die Entwicklung soll schon vorhandene Software (Betriebssystem, Compiler) und E/A-Konzepte (z.B. VME-Bus, P-Bus) berücksichtigen (dadurch konnte die Entwicklung sicher, termin- und kostengerecht durchgeführt werden).

In der Hardware :

- Schnittstelle zur manuellen Eingabe,
- Schnittstelle zur Dateneingabe mit Barcode,
- Kommunikationsfähigkeit mit übergeordnetem System unter Berücksichtigung der Systembelastung auch bei großer Anzahl von BDEs (dadurch wird Polling ausgeschlossen),
- lokale Bearbeitung von Prozeßsignalen zur höheren Systemverfügbarkeit sowie zur schnellen Reaktion,
- modulare Schnittstelle für dezentral anschließbare digitale E/A (die Prozeßsignale werden auf der E/A-Schnittstelle aufgelegt, zum BDE-Terminal führen daher nur wenige Leitungen).

In der Systemsoftware :

- Verwendung eines vorhandenen Betriebssystems,
- Programmierung in höherer Programmiersprache,
- Unterstützung des gewählten Kommunikationsprotokolls mit kurzen Übertragungs- und Antwortzeiten,
- Erkennen lokaler Hardwarefehler (Prozeß-Schnittstelle),
- Absicherung des ganzen Systems gegen lokale Ausfälle von BDE's (ein ausgefallenes Terminal darf nicht das gesamte System blockieren),
- Programme für lokale Diagnose und Ferndiagnose durch das übergeordnete System,
- Warmstart-Fähigkeit nach dem Stromausfall (mit unverändertem Datenbestand),
- geänderte Anwendungssoftware soll vom übergeordneten System an die einzelnen BDEs geladen werden.

In der Anwendungssoftware :

- lokale Benutzerführung unter Berücksichtigung des aktuellen Prozeßzustandes,
- Plausibilitätsprüfungen für Formate und zugelassene Wertebereiche,
- Plausibilitätskontrolle der peripherien Geber anhand des logischen Prozeßabbildes,
- Wartbarkeit und Änderungsfreundlichkeit der Anwendungssoftware.

Nach einer Marktanalyse und Bewertung eigener Kenntnisse entschieden wir uns für den Motorola-Prozessor 68008 /1/ mit RTOS /2, 3/ als Betriebssystem und PEARL /4/ als Programmiersprache. Es wurde geplant, nur notwendige systemspezifische Treiber im 68000-Assembler /5/ zu schreiben und als

RTOS-Erweiterungen zu integrieren. Die Verbindung mit der Peripherie soll auf dem P-Bus-Konzept /2/ basieren (insbesondere aus Kostengründen - die VME-Lösung war zu teuer).

3. Hardwarelösung

Das entwickelte BDE-Terminal wurde Kommissionier-Terminal KT2 benannt, weil es für zwei Kommissionierplätze (Übergabeplätze) vorgesehen wurde. Das Terminal verfügt über :

- einen leistungsstarken Prozessor Motorola 68008,
- ROM- und RAM-Speicher (64k und 64k erweiterbar), wobei der RAM-Speicher batteriegepuffert ist,
- Spannungsversorgung 24V DC,
- eine alphanumerische Tastatur und 12 Funktionstasten,
- ein LCD Display mit 2 Zeilen (insgesamt 80 Zeichen),
- 4 serielle Schnittstellen (2 x RS-232 und 2 x RS-485) unterschiedlich konfigurierbar,
- Anschluß für Periperiebus (P-Bus),
- Barcode-Decoder für alle gängigen Barcodes,
- bis zu 256 digitale Ein- und Ausgänge.

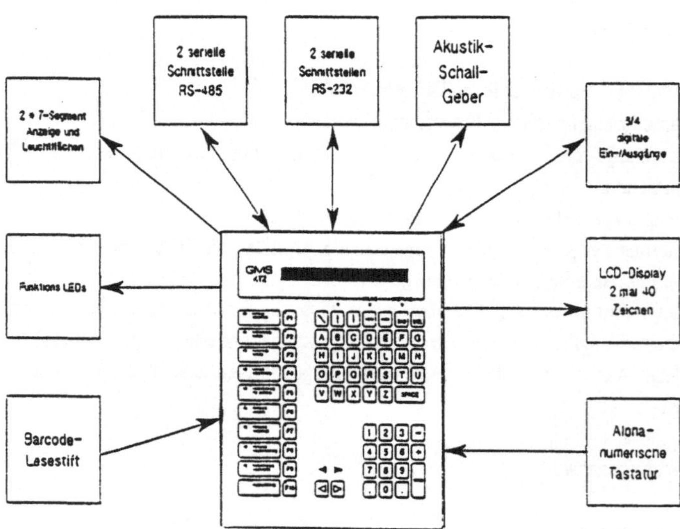

Bild 3 - Konfiguration des BDE-Terminals in einer ausgeführten Anlage

Über die zwei freien RS-232 Schnittstellen wird die Kommunikation mit externen alphanumerischen Terminals ermöglicht (auch PC mit entsprechender Terminalemulation). Sie werden von RTOS unterstützt und erlauben alle interaktiven RTOS-Kommandos. In unserem praktischen Einsatz werden diese Schnittstellen im normalen Betrieb nicht benutzt und sind nur für Inbetriebnahme und Tests des BDE-Terminals vorgesehen.

Die zwei RS-485 Schnittstellen sind für zwei verschiedenen Netzwerk-Anschlüsse vorgesehen (eine für den Ring und eine für den Bus). Um ein ausgeschaltetes oder ausgefallenes Terminal vom Ring zu trennen und dadurch ungestörte Arbeit des restlichen Netzwerk zu garantieren, ist die Ring-Schnittstelle durch ein Relais (Watchdog) abgesichert, das im nicht aktiven Zustand den Ring durchverbindet. Erst mit dauernder Aktivierung des Relais (durch eine Softwareausgabe auf eine bestimmte P-Bus-Adresse) wird das Terminal im Ring aktiv.

4. Softwarefunktionen und Softwarestruktur

Um die Vernetzungsfähigkeit des Terminals zu erreichen mußte neben der Hardwarelösung auch ein entsprechendes Softwarekonzept ausgearbeitet werden. Bei der Vielzahl notwendiger BDE-Terminals kommt der Anbindung an ein übergeordnetes Rechnersystem sehr große Bedeutung zu. Ausgehend von Gesamtsystemen mit mehreren hundert BDE-Terminals scheidet ein Pollingverfahren wegen der langen Reaktionszeiten und der Belastung des übergeordneten Rechners aus. Das Netzwerk basiert daher auf einer Anordnung, die bei Funktionsfähigkeit eines Teilnehmers durch diesen Teilnehmer selbst in einen Ring umkonfiguriert wird. Kennzeichnend für dieses System ist, daß alle Teilnehmer bei Bedarf senden können. Das Polling entfällt, Kollisionen können nicht auftretet. Das System kann mit hohen Datenrate (9600 Baud) störungsfrei betrieben werden. Auf Grund der RS-485-Spezifikation sind maximal 32 Teilnehmer je Ring möglich (die Anzahl der Ringsysteme am übergeordneten Rechner kann frei gewählt werden und ist nur durch die Leistungsfähigkeit des übergeordneten Rechners begrenzt). Die Adreßierung im Netzwerk ist so ausgelegt, daß Nachrichten in Form eines Telegramms zwischen beliebigen Teilnehmer ausgetauscht werden können. Es besteht keine Notwendigkeit einen Teilnehmer als Master zu definieren.

Die vorhandene Möglichkeit freien Telegrammaustausches zwischen beliebigen Teilnehmern wird in diesem praktischen Einsatz nicht benutzt. Es existieren keine Informationsflüsse, die direkt verschiedene Übergabeplätze verbinden. Im Informationsaustausch wird immer die übergeordnete Ebene mit eingezogen. Deshalb haben wir uns entschlossen, daß der Anlagenrechner die Funktion eines Ring-Masters übernimmt und anders als BDE-Terminals auf den Ring zugreift. Der Ring wird vom Anlagenrechner als eine normale RS-232C/V24 betrachtet. Die Telegramme von/zu den BDE-Terminals werden vom Anlagenrechner werden als ganze Einheiten empfangen bzw. gesendet. Es erfolgt keine Weiterleiten von Telegrammen. Die BDE-Terminals arbeiten nach einem anderen Prinzip. Das Terminal fängt nach dem Einschalten sofort mit Empfang an und wartet auf eine bestimmte Start-Kennung. Nach dem Empfang der Start-Kennung wird das Telegramm zeichenweise weitergeleitet und gleichzeitig für weitere Analyse gepuffert. Nach dem Empfang der bestimmten Ende-Kennung (oder Time-out) wird das Weiterleiten bis zur nächsten Start-Kennung eingestellt und der empfangene Puffer untersucht. Wenn die Prüfsumme und die Empfangsadresse stimmen wird der Inhalt der eigenen nächsten Kommunikationsebene übergeben. Dieses Verfahren erlaubt Störungen, die zwischen den Telegrammen auftreten auszufiltern. Beim Einschalten kann es passieren, daß ein Telegramm unterbrochen und nicht weitergeleitet wird (durch das Einschalten des Relais entstehen Störungen). Das wird durch die nächste Kommunikationsebene erkannt, weil alle Telegramme logisch zu quittieren sind. Um zu senden, wartet das BDE-Terminal auf die Ende-Kennung der bereits weitergeleiteten Telegramms und dann sendet sein eigenes Telegramm sofort, wobei die eventuell empfangene Information gleichzeitig gepuffert und sofort nach dem eigenen Telegramm gesendet wird. Das Verfahren ermöglicht hohen Durchsatz. Die Nachrichtenverzögerung beträgt nur etwa 1 Zeichen je Teilnehmer.

Das angewendete Kommunikationsprinzip ist schematisch im Bild 4 dargestellt.

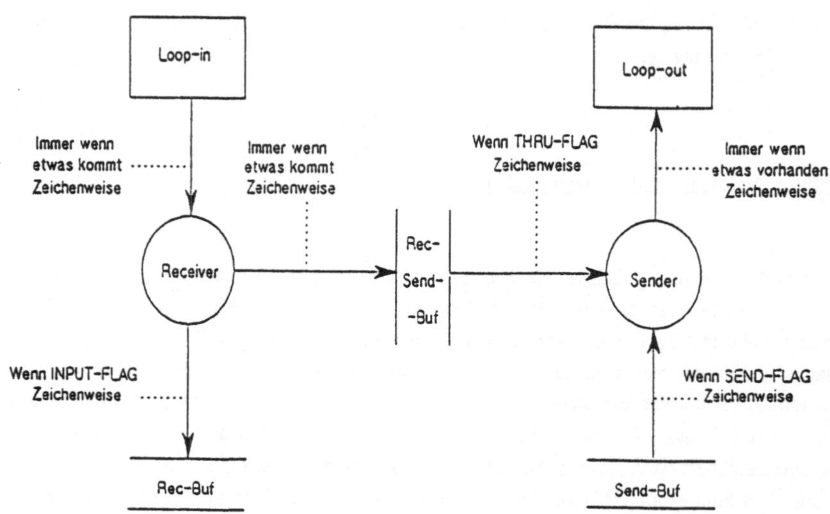

Bild 4 – Empfang, Durchschleusen und Senden von Daten im BDE

Uns ist es nicht gelungen, die entsprechende Kommunikations-Software innerhalb PEARL zu implementieren und das RTOS-DATION-Konzept anzuwenden, weil der erreichte Durchsatz zu niedrig war. Die Ring-Schnittstelle wird mit einem im 68000-Assembler geschriebenen Programm bedient. Das Programm wird beim Systemstart aktiviert und übernimmt vollständig die Kontrolle über dem DUART-Baustein /4/ (serielle Schnittstelle). Das Programm soll im Supervisor-Modus des Prozessors arbeiten, deshalb wurde seine softwaremäßige Schnittstelle mit einigen Traps realisiert. Das Programm kann als sog. RTOS-Scheibe in das Betriebssystem integriert werden. Das Modell des Treibers wurde mit Hilfe eines Zustandsgraphen definiert und entsprechend programmiert.

Die Struktur der Systemsoftware und die Informationsflüsse innerhalb der Systemsoftware werden im Bild 5 dargestellt. Die Verbindung mit der Anwendungssoftware ist nicht dargestellt.

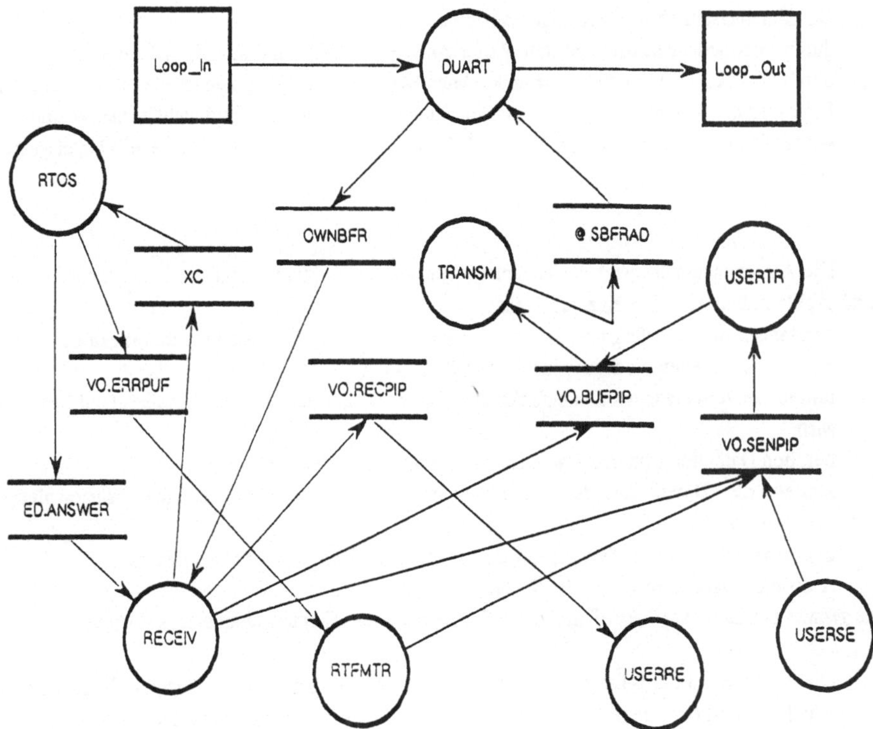

Bild 5 - Prozesse und Informationsflüsse im BDE-Terminal (Systemsoftware)

Die dargestellten Prozesse wurden als RTOS-Tasks implementiert (mit der Ausnahme vom Netzwerk-Treiber). Sie unterstützen folgende Funktionen :

- DUART sorgt für die Kommunikation mit dem Netzwerk, empfängt und sendet Telegramme,
- RECEIV analisiert und quittiert empfangene Telegramme, verteilt sie zwischen anderen Prozesse, unterstützt die Kommunikation mit dem RTOS (virtuell Terminal-Schnittstelle),
- RTFMTR empfängt die RTOS-Fehlermeldungen (das BDE-Terminal verfügt im normalen Betrieb über keine Operator-Konsole) und sendet sie als Telegramme,
- TRANSM bedient die Telgramm-Warteschlange, übergibt die Telegramme einzeln dem Treiber,
- USERRE übernimmt die Daten, die für die Anwendungssoftware bestimmt sind,
- USERSE übernimmt die Daten von der Anwendungssoftware,
- USERTR übernimmt die Daten zum Senden aus der Warteschlange, baut daraus Telegramme und steckt sie in die Telegramm-Warteschlange ein..

Der Datenaustausch wird wie folgt realisiert :
- durch verschiedene selbst verwaltete Puffer zwischen DUART und RECEIV/TRANSM,
- durch vom RTOS unterstützte virtuelle Datenstationen (mit VO gekennzeichnet) zwischen anderen Prozessen der Systemsoftware und auch von/zu der Anwendung (im Bild nicht berücksichtigt),
- durch Kommando-Datenstation XC mit RTOS, wobei die Antwort auf eine ED-Datei gespeichert wird.

Die Anwendungssoftware realisiert anlagenspezifische Funktionen und beinhaltet mehrere Tasks die parallel ablauffähig sind. Die einzelne Tasks :
- empfangen die Prozeßsignale, bearbeiten sie, bestimmen die Zustandsänderungen im technischen Prozeß und bereiten die Steuersignale zum Prozeß vor,
- führen die verschiedenen Dialogfunktionen durch, wobei der aktuelle Prozeßzustand berücksichtigt wird,
- bereiten Daten für zyklische Statustelegramme und Dialogtelegramme vor,
- senden die Daten für den Anlagenrechner durch die entsprechende Warteschlange zur Systemsoftware,
- übernehmen die von der Systemsoftware empfangenen Daten vom Anlagenrechner,
- realisieren Datenanzeige zum Bediener.
Die genaue Beschreibung dieser Funktionen überschreitet das Themengebiet dieses Vortrages.

Die Software des BDE-Terminals ist in Form von zwei weitgehend unabhängigen Einheiten realisiert. Die Aufteilung entspricht dem oben genannten. Das RTOS-Betriebssystem und seine systemspezifischen Erweiterungen sind im PROM-Speicher enthalten. Die anwendungsspezifische Softwareteile werden in RAM-Speicher vom übergeordneten Rechner geladen.

Als Grundmodell für die Softwareerstellung haben wir das Zustandsmodell eingesetzt. Das Modell wurde benutzt um mögliche Abläufe darzustellen :
- als Grundlage für den Ring-Treiber in der Systemsoftware,
- als Ablaufbeschreibung für Übergabeplätze in der Anwendungssoftware.
Die spezifizierte Beschreibung wurde direkt in die Software übernommen.

5. Industrieller Einsatz

Das BDE-Terminal befindet sich seit Anfang des Jahres 1990 im industriellen Einsatz. Es wurde in einem Lager in einem hierarchischen Steuerungssystem integriert. Das System verfügt über 128 Terminals, die in mehreren (acht) Netzen gekoppelt sind und mit dem übergeordneten Rechner (microVAX mit VMS-Betriebssystem) kommunizieren.

6. Probleme, Erfahrungen aus der Entwicklung und Inbetriebnahme

Die gewählte PEARL-Programmiersprache hat sich sehr gut bewährt. Mit der Ausnahme einiger hardwarenaher Software konnten alle Programme in PEARL geschrieben werden. Die Notwendigkeit auch Assembler zu benutzen ist entstanden :

- auf Grund des gewählten Kommunikationsprinzips (im Ring) und gewünschten hohen Durchsatz,
- durch Initialisierungsprobleme für einige statische Speicherstrukturen (z.B. Scan-Code-Tabelle).

Zur Anwendung von RTOS kann gesagt werden :

- die notwendige RTOS-Erweiterungen sind mit etwas Erfahrung einfach zu entwickeln und durch das RTOS-Scheiben-Konzept relativ einfach integrierbar,
- die RTOS-Leistung und Reaktionszeit sind für unsere Anwendung vollkommen ausreichend,
- die Power-Fail- und Warmstart-Funktionen waren nicht vorhanden, es besteht immer eine Möglichkeit der Warmstart durch ein Power-Fail während des Warmstarts zu zerstören und automatisch zum Kaltstart zu übergehen,
- das RTOS-Betriebssystem kann wesentlich umfangreichere Aufgabe übernehmen, von uns wurden nicht alle Möglichkeiten vollständig ausgenutzt (RTOS ist auch parallel zu unserem Projekt gewachsen und wurde seit Herbst 1988 sehr stark erweitert).

Bei der Inbetriebnahme des Systems sind auch einige Probleme entstanden :

- Bei kurzen Stromausfällen sind Probleme mit intelligenten Bausteinen aufgetreten. Der Reset-Logik ist besondere Aufmerksamkeit zu widmen, da sonst durch falsche Arbeitsweise bei der Initialisierung der Rechnerspeicher zerstören werden kann.
- Es ist uns nicht gelungen in einer einfachen Weise die saubere Trennung zwischen Programmen und Daten für RAM-Software zu erreichen (sie ist nur für PROM-Software unterstützt), deshalb waren auch durch Störungen entstandene Softwaremodifikationen schwierig zu entdecken.
- Der Prozessor hat keinen hardwaremäßigen Speicherschutz, erlaubt deshalb keine Zugriffskontrolle. Deshalb wird der Komfort einer höheren Programmiersprache und die vorausgesetzte Sicherheit vermindert (einige Zeiger-Fehler oder Speicherzerstörung können unvorhersehbare Konsequenzen haben),
- Wir hofften, keine Änderungen in der PROM-Software in der Anlage durchführen zu müssen (weil sie bei mehr als 100 BDE-Terminals sehr aufwendig sind). Leider mußten wir die PROMs zweimal wegen zusätzlicher Funktionen für Diagnose und wegen Power-Fail-Problemen umtauschen.

Bei der großen Anzahl der BDE-Terminals hat sich insbesondere die Möglichkeit bewährt, die Anwendungssoftware von der VAX durch den Ring zu Laden. Die ganze Anlage kann in etwa 2 Stunden mit der neuen Software-Version geladen werden, der normale Betrieb der Anlage braucht nicht zu unterbrochen werden.

Die Realisierung eines virtuellen Terminals durch der Ring war auch vorteilhaft. Das virtuelle Terminal darf jedoch nur mit Vorsicht und nur durch erfahrene Benutzer (nur Wartungszwecke) gebraucht werden. Die Gefahr besteht daran, daß durch falsche RTOS-Kommandos (z.B. Speicherauszug von nicht existierenden Adressen), die durch eine XC-Datenstation übergeben wurden, stürzt die entsprechende Task mit Motorola BUS-Error ab.

Die zweite Netzwerk-Schnittstelle (Bus) die für Diagnosezwecke vorgesehen ist, wurde nicht benutzt. Deshalb ist der Ring in Fall eines Softwarefehler nicht automatisch rekonfigurierbar (obwohl Diagnose am Anlagerechner beim Ausfall eines BDE sehr einfach ist). Die früher beschriebene automatische Trennung des BDE-Terminals vom Ring gibt auch keine hundertprozentige Absicherung - die Task, die das Relais bedient, kann laufen und das Relais zyklisch aktivieren, obwohl z.B. die anderen Kommunikations-Tasks abgestürzt sind, oder auf sog. Kommunikations-Workspace (CWS) warten und dadurch blockiert sind. Deshalb haben wir einige Ring-Ausfälle beobachtet, die jedoch einfach zu beheben wurden.

Der Einsatz der Zustandsgraphenmodelle für den Softwareentwurf hat sich als sehr gut bewährt weil :
- die aus dem Modell entstandene Software eine klare Struktur hat und dadurch änderungsfreundlich und wartbar ist,
- die Zustände des technischen Prozesses können zu den übergeordneten Ebenen gesendet und dort visualisiert werden (ist vom großem Vorteil für Benutzer).

Der Entwicklungsaufwand für BDE-Terminal kann wie folgt angegeben werden :
- für Hardware - 12 MannMonate,
- für Software - 12 MannMonate.

Literatur

/1/ J.Assenbaum, A.Hadler, B.Kroll, B.Schöfer "Zwergenaufstand" - eine Reihe von Aufsätzen in C't, Hefte 2 bis 7, 1987
/2/ "RTOS/PEARL Integriertes Echtzeit-Multitasking-Programmiersystem" Heinz Heise Verlag 1989
/3/ K.Koerth, C. Weitz "Echzeit-Multitasking mit RTOS/PEARL" - eine Reihe von AUfsätzen in C't, Hefte 6/1987 bis 3/1988
/4/ W.Werum, H.Windauer "PEARL - Prozess and Experiment Automation Realtime Language" Vieweg 1978
/5/ K.Kief "Assembler-programmierung mit dem M68000" Vieweg & Sohn 1988

Regelung eines Tiefsee-Hammers
mit Hilfe eines PEARL-Rechner-Systems

Werner Schulze

esd electronic system design gmbh
Vahrenwalderstraße 7, D-3000 Hannover 1
Telefon: (0511) 3563 380

Zusammenfassung:

In Form eines Applikationsberichtes wird der Einsatz eines VMEbus-Rechner-Verbunds in der Offshore-Technik geschildert. Für den erstmaligen Versuch, Pfähle in einer Tiefe von 1000 m in den Meeresboden zu rammen, mußte neben dem hohen technischen Aufwand in der mechanischen Ausführung auch für die Automatisierungskomponenten Neuland betreten werden. Unter Erläuterung einiger spezieller Probleme wird die realisierte Hardware-Lösung kurz vorgestellt. Weiter wird auf die Anforderungen aus dem Echtzeitbereich und deren Umsetzung in äußerst praktikable und flexible Software-Konzepte auf der Basis eines PEARL/RTOS-UH Multi-Tasking-Systems eingegangen. Der Bericht schließt mit konkreten Problemlösungen für einige Störfälle, die letztlich bei Pilotprojekten unvermeidbar sind. Die Operation wurde als voller Erfolg mit der Niederbringung eines 100 m langen Pfahls bei 1000 m Wassertiefe im Dezember 1989 abgeschlossen.

1. Einleitung

Bereits in den letzten Jahren hat sich in der Offshore-Technik bei der Einrammung von Pfählen für die Befestigung von Bohrinseln oder Unterwasserrohrleitungen in Bezug auf die Automatisierung der Trend abgezeichnet, an Stelle von SPS-gestützten Steuerungen kompakte Prozeßrechner direkt vor Ort einzusetzen. Der Vorteil des Rechnereinsatzes liegt vor allem in der Möglichkeit, durch spezielle Algorithmen eine Regelung des Hammers für die Rammung vorzunehmen, die nicht nur versucht, stets auftretende Störungen auszugleichen, sondern sich auch weitgehend automatisch auf wechselnde, bzw. veränderliche Bodenstrukturen einzustellen. Der Gewinn liegt dann in der gegenüber der konventionellen Rammung wesentlich schnelleren Niederbringung des Pfahles bei gleichzeitig verbesserter Sicherheit bezüglich der Verankerung des Pfahles im Meeresboden.

Als weiterer Vorteil hat sich die Aufzeichnung und Protokollierung relevanter Meßdaten und daraus abgeleiteter Prozeßgrößen wie etwa Schlagenergie und Eindringgeschwindigkeit während des Rammvorganges, die praktisch als Abfallprodukt der Automatisierung auftreten, herausgestellt. Insbesondere vom internationalen Versicherungswesen werden diese Daten mittlerweile sogar gefordert, um auch im Nachhinein eine Rammung beurteilen zu können und die nicht gerade geringen Risiken für den festzusetzenden Pfahl zu übernehmen.

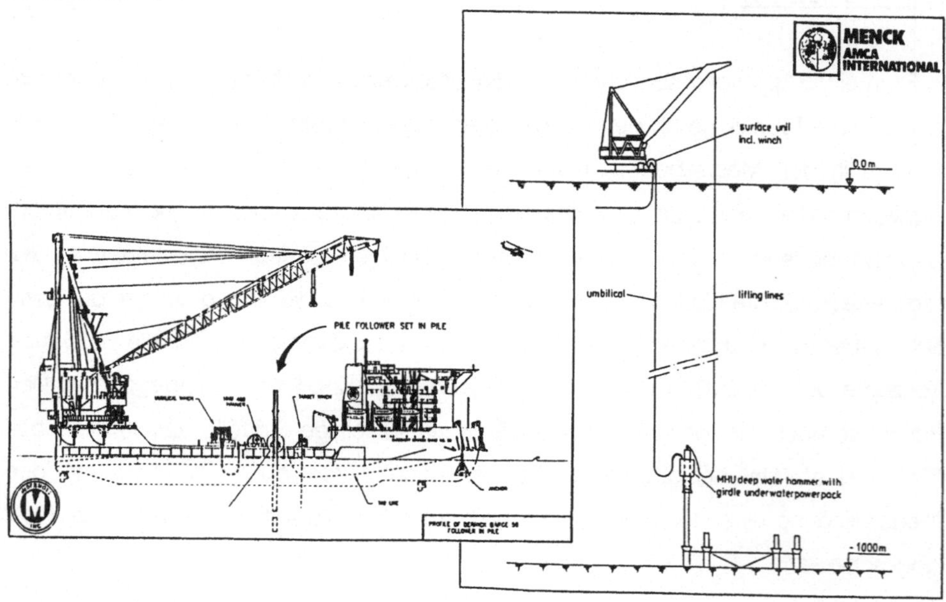

Bild 1:   Einsatzskizze für den Tiefsee-Hammer MHU 400T und Detailbild für das Bohrschiff mit Absenkeinrichtung für Hammer und Pfahl.

2. Stabile Hardware und Software

Als ein wesentlicher Aspekt für den Rechnereinsatz muß bei den immensen Kosten im Offshore-Bereich die hochgradige Zuverlässigkeit der Hardware und Software der Automatisierungskomponenten gelten. Hier hat sich die in der Offshore-Pfahl-Technik führende Firma MENCK aus Hamburg im Hardware-Bereich für das VMEbus-System entschieden.

Als CPU-Karten wurden dabei VMEbus-Boards von FORCE, München, auf der Basis MC 68020/68882 gewählt, während als Prozeß-I/O-Karten für die analogen, digitalen und seriellen Signale VMEbus-Boards von **esd** aus Hannover zum Einsatz kamen, die sämtlich eine galvanische Trennung der Prozeßsignale gegenüber dem VMEbus-System aufweisen.

Hinsichtlich der Software entschied man sich für den Einsatz von PEARL, einer nach DIN 66 253 genormten Hochsprache für Echtzeitprobleme, auf der Basis des Multi-Tasking-Betriebssystems RTOS-UH, da von vorn herein abzusehen war, daß aus dem Echt-zeitbetrieb erhebliche Anforderungen an das System gestellt werden würden. Die Nutzer-Programme für Regelung und Daten-Archivierung wurden ausschließlich vom Anwender selbst und auf Hochsprachen-Niveau geschrieben, lediglich die Treiber für die Prozeß-I/O-Karten wurden von **esd** standardmäßig gestellt.

Mittlerweile sind mehrere dieser Systeme bei diversen Bohrinseln und Bohrschiffen zur vollsten Zufriedenheit im Betrieb, so unter anderem in den Ekofisk- und Veslefrik-Feldern der Nordsee sowie in der Adria, vor Kalifornien und bei Südafrika. Die Wassertiefen liegen dabei üblicherweise zwischen 80 m und maximal 365 m.

3. Extreme Anforderungen

Aufbauend auf diesen Erfahrungen sollte nunmehr einer Herausforderung begegnet werden, die eine Rammung in 1000 m Wassertiefe vorsah. Ein wahrer Meilenstein, wenn man bedenkt, daß der Tiefenrekord bis zu diesem Zeitpunkt bei 540 m lag und nur unter größten Schwierigkeiten überhaupt erreichbar war.

Bei allen Beteiligten bestand Eingikeit darin, daß mit diesem Projekt auf vielen Gebieten zugleich Neuland betreten werden mußte. Deshalb wurde ein Pilotprojekt zwischen diversen Ölgesellschaften und Offshore-Betreibern vereinbart, das die erfolgreiche Nieder-bringung eines 100 m langen Pfahles in 1000 m Wassertiefe als Testfall vorsah.

Das Hauptproblem besteht in der großen Entfernung zwischen dem Einsatzort des hydraulisch arbeitenden Hammers am Meeeresboden und der 1000 m darüber befindlichen Kontrollebene an Deck des Bohrschiffes (Bild 1). Über diese riesige Strecke ist nicht nur die Energie- und Luft-Versorgung des Hammers sicherzustellen, sondern ebenso der reibungslose Datentransfer.

Kenndaten:

Gewicht des Hammers:	93.4 t
Länge:	16.7 m
max.Breite:	4.6 m
max. Schlagenergie:	400 kNm
min. Schlagenergie:	40 kNm
Spannungsversorgung:	3.3 kV
max. Leistung:	420 kW
Arbeitsdruck:	210 bar
Gewicht Winde u. Umbilical:	144.5 m
Länge:	1200 m
Duchmesser Umbilical:	140 mm

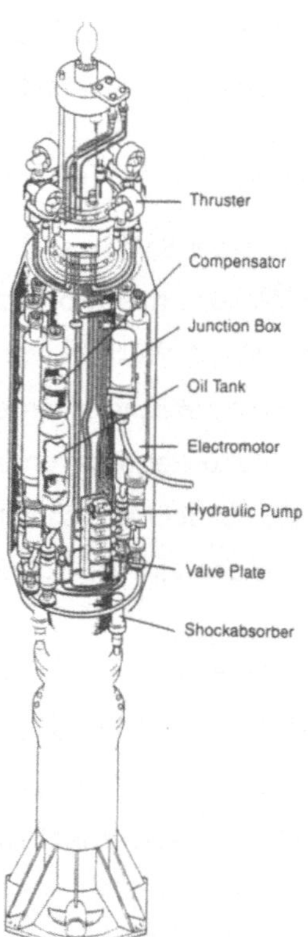

<u>Bild 2:</u>   Skizze des MENCK Tiefsee-Hammers MHU 400 T mit integriertem Unterwasser-Power-Pack sowie Kenndaten des Hammers.

Um ein einfaches Handling des Hammers auf See zu ermöglichen, wurde der Hammer in einer Neukonstruktion so aufgebaut, daß er mit dem Unterwasser-Power-Pack eine Einheit bildet (Bild 2). Das Power-Pack setzt die elektrische Energie mit Hilfe von vier einzeln schaltbaren Motoren über Hydraulikpumpen in hydraulische Energie um, die dann bei einem Arbeitsdruck von ca. 200 bar den Hammer auf- oder abwärts beschleunigt. Das Power-Pack ist wie ein Gürtel um den Hammer montiert und stützt sich über Schockabsorber gegenüber dem Hammerkörper ab.

Schließlich wurde aus Effizienz- und Redundanzgründen ein Konzept aufgestellt, das die Teilbarkeit der Gesamtanlage in die Komponenten Hammer mit Unterwasser-Power-Pack für die Öldruck-Versorgung sowie Schaltanlage und Steuerstand durch die Schaffung eines verteilten Rechnernetzes mit einer direkten Zuordnung zu eben diesen Einheiten berücksichtigte (Bild 3). Das aber hieß, daß erstmals ein VMEbus-System in 1000 m Wassertiefe, dazu noch direkt auf einem mit 400 kNm Schlagenergie arbeitendem Hammer, zum Einsatz kommen sollte.

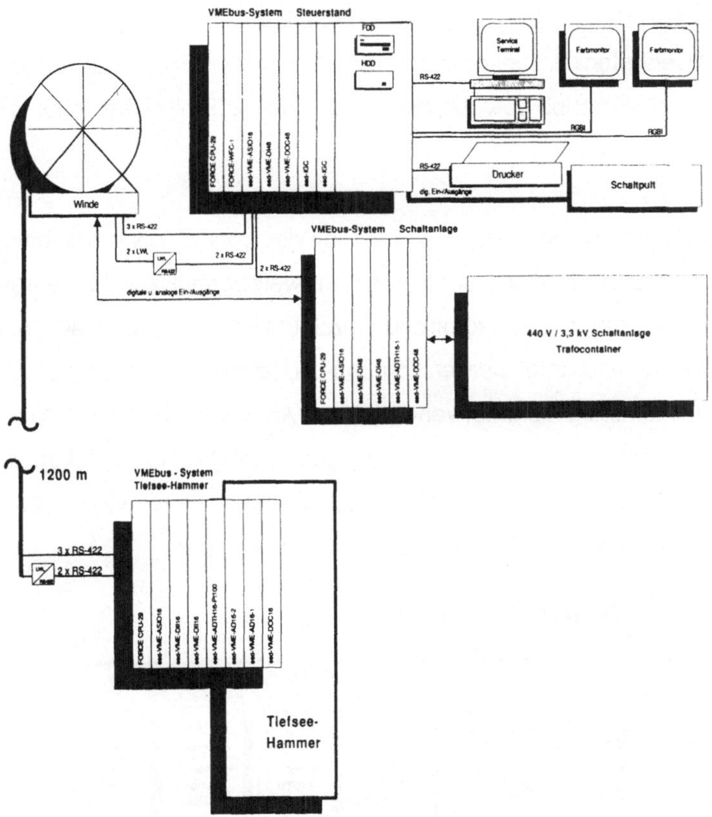

<u>Bild 3:</u>   Rechnerkonzept für die Regelung des Tiefsee-Hammers MHU 400 T mit den drei Komponenten für Tiefsee-Hammer, Steuerstand und Schaltanlage.

Die Verbindung zwischen Hammer und Schiff wird durch eine sogenannte Nabelschnur (UMBILACIL) gehalten, in der dem am Hammer montierten Unterwasser-Power-Pack über 5 Adern-Systeme jeweils 3.3 kV / 60 Hz mit max. 750 kVA zugeführt werden. Über eingelassene Pressluftschläuche wird der Hammer von einem an Deck stehenden

Kompressor mit 200 KW Leistung bei einem Druck von ca. 100 bar mi. Luft versorgt, deren Überdruck im Arbeitsvolumen des Hammers bei der vorgesehenen Tiefe aber nur noch 1 bar beträgt.

4. Der Rechner am Hammer

Der VMEbus-Rechner direkt am Hammer (Bild 4) ist in einer druckfesten Stahltrommel montiert, die über zusätzliche Stahlfedern in das Power-Pack eingehängt wurde, um die Sckockbelastung wenigstens einigermaßen erträglich zu halten. Für den Systemaufbau wurde ein Standard 19"-Einschub gewählt, der lediglich auf 42 TE mit einem 9 Slot VMEbus gekürzt wurde.

Auch die VMEbus-Boards waren Standardkarten "von der Stange", ohne daß etwa MIL-Versionen oder besonders selektierte Boards bemüht werden mußten. Sehr entgegenkommend war hierbei natürlich, daß sämtliche Prozeß-I/O-Verbindugen direkt über den P2-Stecker des VMEbus verdrahtet werden konnten und somit mechanisch aufwendige Frontplatten-Steckerversionen nicht notwendig waren. Als äußerst vorteilhaft erwies sich auch, daß die galvanische Trennung der Prozeßsignale direkt durch Optokoppler auf dem Board erfolgt und damit - außer für die RS-422/LWL-Umsetzung -keine zusätzlichen Befestigungsprobleme für weitere Vorverarbeitungskarten anfielen.

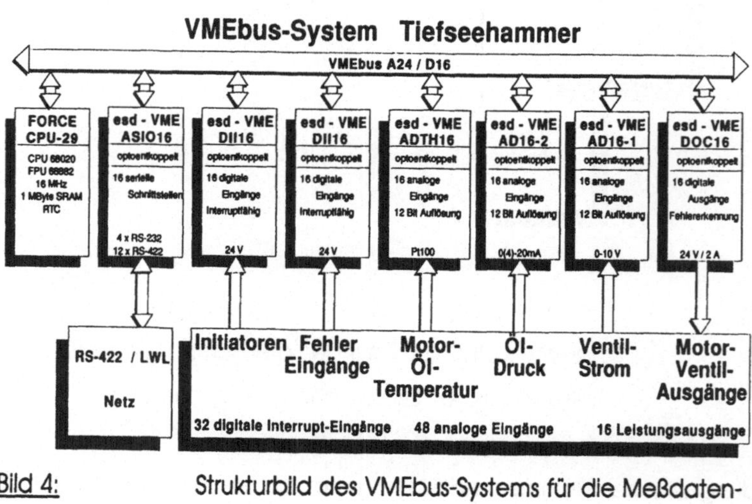

Bild 4: Strukturbild des VMEbus-Systems für die Meßdaten-Erfassung und Regelung direkt am Tiefsee-Hammer.

Der Hammer-Rechner hat vornehmlich die Aufgabe, die Umsteuerzeiten für die Hydraulikventile aus den vorgegebenen Energiewerten sowie der tatsächlichen Beschleunigung des Hammers zu berechnen. Dabei wird dem Rechner die Ist-Position des Hammers per Interrupt von Initiatoren gemeldet, die durch nachgeschaltete Hardware-Timer bis in den Mikro-Sekunden Bereich aufgelöst werden können. Auf diese Weise kann die momentane Schlagenergie des Hammers sehr genau berechnet werden und steht damit als die eigentliche Regelgröße des Prozesses zur Verfügung.

Die Ermittlungsweise dieser Regelgröße ist als durchaus kritisch anzusehen. Fehler, die durch ein unsauberes Interrupt-Handling von Hardware oder Software hervorgerufen werden, führen sofort zu Fehlberechnungen der Schlagenergie und lassen den Hammer nicht nur unökonomisch arbeiten, sondern unter Umständen seine Sollenergie niemals erreichen. Hier ist - korrekte VMEbus-Hardware vorausgesetzt - vor allem ein sauber definiertes Echzeitverhalten der Software gefordert. Wohlbekannte Betriebssysteme mit lediglich echzeit-ähnlichen Eigenschaften würden das Projekt unweigerlich scheitern lassen.

Eine sehr kritische Passage stellt das Durchlaufen des Hammers im oberen Umkehrpunkt dar. Während eine falsche Steuerzeit im unteren Umkehrpunkt als Fehlanpassung gewertet werden kann, bei der es nicht gelingt, die gesamte Schlagenergie des Rammgewichtes an den Pfahl weiterzuleiten, können Steuerfehler beim Aufsteigen des Rammgewichtes zu dauerhaften Schäden führen. Im schlimmsten Fall schlägt das Rammgewicht gegen die obere Abschlußplatte und zerstört den Hammer.

Dennoch ist es nötig, den Hammer bei jedem Schlag immer wieder in die Nähe dieses kritischen Punktes zu bringen, um durch Ausnutzung des maximalen Hubes auch die größtmögliche Schlagenergie in den Pfahl zu bringen. Softwaremäßig wurde dieses Problem entsprechend den Regeln der Parallelverarbeitung durch Schaffung jeweils einer zu einem Ereignis zugeordneten Task angegangen. Die Aktivierung dieser Tasks erfolgt dann durch das Eintreffen eben dieses Ereignisses. Zusätzliche über Zeitbedingungen eingeplante Überwachungs-Tasks sorgen dafür, daß auch bei ausbleibenden Meldungen von z.B. defekten Initiatoren, ein sicherer Steuerablauf gewährleistet bleibt.

Auf einem niedrigeren Prioritätsniveau werden über weitere Tasks analoge Motor-, Öl- und Hammer-Daten zyklisch abgefragt. Bei Überschreitung von Grenzwerten werden automatisch Warn- oder Alarm-Behandlungen eingeleitet. Das gesamte Tasking-Konzept ist so aufgebaut, daß auch für den Extremfall einer vollständigen Kommunikationsunterbrechung mit dem Steuerstand der Hammer im Stand-Alone-Betrieb weitergeregelt wird. Selbst nach Ausfall der Stromversorgung wird bei Wiederkehr automatisch mit den zuletzt aktuellen Parametern der Rammbetrieb in der Tiefe wiederaufgenommen. Das heißt, in einer derartigen Notlage reduziert sich der Einfluß vom Bohrschiff aus auf das Ein- und Aussschalten der Steuer- und Motorspannungen - ansonsten arbeitet der Hammer vollkommen autark und gewährleistet die erfolgreiche Einrammung des Pfahles.

Eine Besonderheit stellen die Ermittlungen der Verläufe von Vorlauf- und Rücklaufdruck dar, die wegen ihrer kurzen Abtastzeit von lediglich 50 ms durch separate Tasks auf mittlerem Prioritätsniveau erfaßt werden.

5. Der Rechnerverbund

Im Normalfall werden alle direkt vor Ort bestimmten Daten über einen anonymen Leitungsverbund von drei RS-422 Leitungen in Kupfertechnik und zwei Lichtwellenleitern (LWL) mit Baudraten von jeweils 38400 Baud über die 1200 m Strecke dem Master-Rechner im Steuerstand zugespielt (siehe auch Bild 3). Ebenso können neue Kommandos oder Parameter vom Master empfangen werden.

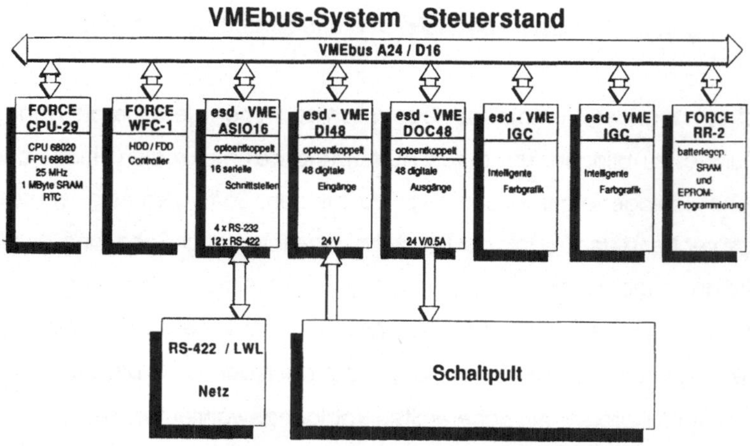

<u>Bild 5:</u>   Strukturbild des VNMEbus-Systems im Steuerstand.

Jede Leitung wird durch Tasks, die jeweils auf den beiden Rechnerseiten ein Paar bilden, betreut. Bei Leitungsausfall oder auch kurzzeitigen Störungen werden die Daten selbständig an intakte Kommunikations-Paare weitergereicht. Bei totalem Kommunikationsverlust fällt der Hammer-Rechner in den Autark-Modus, die Daten können in einem batteriegepufferten SRAM gespeichert werden und nach Aufholen des Hammers für Diagnose-Zwecke ausgelesen werden.

Der Steuerstand-Rechner (Bild 5) dient im wesentlichen als Bedieninterface. Über zwei Farbgrafikschirme wird die Bedienmannschaft detailliert über den Prozeßzustand informiert. Die Druckverläufe im Hammer sowie Hubhöhe und Geschwindigkeit des Hammers werden in oszilloskop-ähnlicher Weise mit einer Zeitauflösung von 50 ms dargestellt. Alle langsameren Prozeßvariablen werden in separaten Fenstern als Linienschrieb aktualisiert oder in Form von Bargraphen bilanziert.

Der Steuereingriff erfolgt über ein klassisches Schaltpult mit Drucktastern, die auch von ölgetränkten Lederhandschuhen bedient werden können, und Lampen für Rück- und Fehlermeldungen, die auch durch die salzverkrustete Schutzbrille wahrgenommen werden. Komplexere Abläufe und Parameter-Änderungen werden über ein festeingebautes Datenterminal vorgenommen.

Für die Kontrolle der 440 V / 3.3 kV Schaltanlage wurde aus Redundanzgründen ein separater VMEbus-Rechner vorgesehen (Bild 6). Ähnlich wie der Hammer-Rechner, kommuniziert er mit dem Steuerstand-Rechner über zwei serielle RS-422-Leitungen von ca. 40 m Länge bei 38400 Baud. Über insgesamt 130 digitale und 15 analoge Kanäle obliegt ihm die Ablaufsteuerung und Überwachung für das Ein- und Ausschalten der Hydraulik-Motoren in der Tiefe.

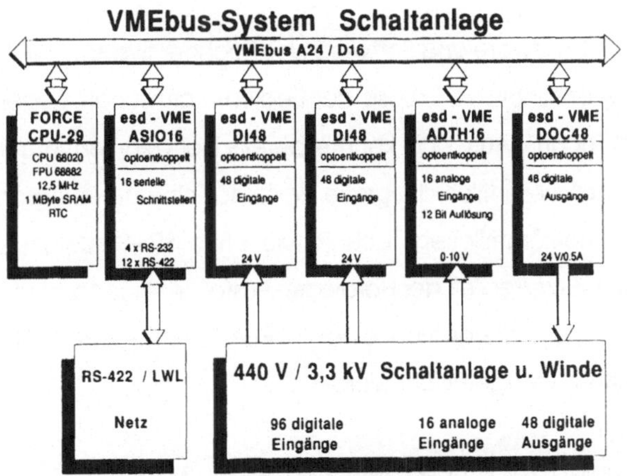

Bild 6:   Strukturbild des VMEbus-Systems für Schaltanlage und Winde.

Die PEARL-Software-Pakete für die drei Rechner haben das folgende Mengengerüst:

Anzahl	Steuerstand-Rechner	Hammer-Rechner	Schaltanlagen-Rechner
Moduln	9	5	3
Tasks	68	62	38
Quellzeilen	7500	4800	2100

6. Inbetriebnahme

Ein erster Trockentest des kompletten Systems wurde im Sommer 1989 in der Ostsee oberhalb der Wasserlinie vorgenommen. Der Test fiel bei einer gemessenen Schlagenergie von 455 kNm zur vollen Zufriedenheit aus. Es wurden in mehreren Sessions insgesamt 11000 Schläge ausgeführt. Weder in der Hardware noch in der Software der VMEbus-Rechner traten Ausfälle auf, so daß die gesamte Anlage sofort für die Verschiffung in die USA eingeschweißt wurde.

Der abschließende Test wurde im Dezember 1989 im Golf von Mexiko auf einem Bohrschiff unter den Augen von Abgesandten diverser Ölgesellschaften in einer Wassertiefe von exakt 1000 m mit einem ca. 100 m langem Pfahl vorgenommen. Der Test war ein voller Erfolg, der Pfahl wurde vollständig bis 1,5 m unter den Meeresboden gerammt.

Auch hier haben Hardware und Software des Rechner-Systems zur vollsten Zufriedenheit gearbeitet. Eine anfängliche Skepsis der Bedien-Mannschaft gegenüber dem ihr groß erscheinenden Technik-Aufwand konnte bereits bei den Testvorbereitungen aufgelöst werden. Hierzu hat wesentlich die große Flexibilität eines frei-programierbaren Rechners gegenüber der herkömmlichen Technik wie etwa SPS-Steuerungen beigetragen. So konnte innerhalb kürzester Zeit der Hardware-Ausfall eines separaten Druckreglers für den Luftkompressor durch Intergration einer in PEARL formulierten PID-Regler-Task in das Steuerstand-System wettgemacht werden.

Diese hohe Flexibilität wird neben dem modularen VMEbus-Konzept vor allem durch das PEARL/RTOS-UH-System gewährleistet und läßt sich auch durch den folgenden Vorfall eindrucksvoll demonstrieren: Bei dem letzten Check unmittelbar vor Niederbringung des Hammers trat bei Anlegen der Motorspannung ein Kurzschluß in einer der 3.3 kV Mittelspannungsleitungen auf. Durch Überschlag auf die anderen Leitungsstränge im Umbilical wurden sämtliche RS-422 Treiber und Empfänger auf beiden Seiten des Umbilicals zerstört. Allerdings konnte der Schaden durch die strikte Einhaltung der galvanischen Trennung begrenzt werden und durch einfaches Austauschen der seriellen VMEbus-Interfaceboards behoben werden. Die VMEbus-Systeme selbst wurden auf der Rechner-Seite nicht beeinträchtigt.

Im Umbilical wurde der defekte 3.3 kV Leitungsstrang durch Umrangieren auf einen Reservestrang ersetzt. Da aber die Zuverlässigkeit des gesmten Umbilical nunmehr mit weniger als 100 % angesetzt werden mußte, sollte durch Software-Änderung der Autarkie-Character des Hammer-Rechners noch weiter verstärkt werden, so daß auf jeden Fall, auch bei Ausfall aller Verbindungen bereits während der Absenkphase, der Rammvorgang im Notfall selbständig eingeleitet und erfolgreich beendet werden sollte.

Für diese Operation standen insgesamt sieben Stunden zur Verfügung, in der die Programmänderungen auf dem Steuerstand-Rechner vorgenommen wurden, mittels einer Simulation im trocken laufenden Hammer verifiziert wurden und ein kompletter Satz von 4 EPROMs mit der neuen PEARL-Anwender-Software gebrannt wurde. Es spricht für die Handhabarkeit des Systems, das dieses Vorhaben gelang und damit einer der Grundsteine für den Erfolg der gesamten Operation gelegt werden konnte.

<u>Literatur:</u>

W.Gerth   PEARL/RTOS_UH, Benutzer-Handbuch, Heise-Verlag, Hannover, 1988.

esd   VMEbus-Mappe, esd electronic system design gmbh, Hannover, 1990.

Menck   Report on DEEPWATER PILE DRIVING TEST, partly confidential, Menck GmbH, Hamburg, 1990.

Neue Implementierungswege mit PEARL 90

Manfred Warzawa

Erwin Kneuer

Werum Datenverarbeitungssysteme GmbH
Erbstorfer Landstraße 14
2120 Lüneburg
Tel. 04131/8900-0

Zusammenfassung

Die unter dem Arbeitstitel PEARL 90 geführte Diskussion über die Weiterentwicklung der Echtzeit-Programmiersprache PEARL ist weitgehend abgeschlossen [Stieger 89]. Die meisten Spracherweiterungen ergaben sich direkt aus den Anforderungen und Erfahrungen der Praxis vieler PEARL-Projekte.

Die Firma Werum --- Anbieter und selbst intensiver Nutzer eines PEARL-Programmiersystems --- entschloß sich, die Spracherweiterungen von PEARL 90 zu implementieren. Dieser Beitrag berichtet über die Wege, die bei der Neuentwicklung des PEARL-Compilers beschritten wurden.

1. Einleitung

Vor mehr als 10 Jahren entwickelte die Firma Werum ein portables PEARL-Programmiersystem (Compiler und Laufzeitsystem), das bis heute auf mehr als zwanzig der unterschiedlichsten Rechnersysteme portiert wurde. Mit der Konkretisierung von PEARL 90 [Stieger 89] und der Entscheidung von Werum diese Spracherweiterungen zu implementieren, stellte sich die Frage einer Weiter- oder Neuentwicklung unseres PEARL-Programmiersystems.

Die Grundstrukturen des Compilers ergaben sich aus dem damaligen Stand der Technik: Erstes Zielsystem war ein 16-Bit Rechner mit maximal 64 KB Hauptspeicher. Die implementierten Algorithmen erwiesen sich bis heute als sehr effizient, allerdings bedingen selbst leichte Änderungen der zu akzeptierenden Quellsprache einen größeren Aufwand zur Anpassung der immerhin 15 Compilerpasse. Einen ähnlich hohen Aufwand würde eine Erweiterung von dem *antiken* 64 KB Adreßraum auf die heute weit verbreiteten 32 Bit Architekturen erfordern, der --- soweit die Hardware und das Betriebssystem des Zielrechners dies ermöglichen --- von unserem Compiler unterstützt werden soll.

Wir entschieden uns daher für eine Neuentwicklung des PEARL-Compilers und eine Weiterentwicklung der portablen Laufzeitkomponenten für das PEARL-Tasking und die Ein-/Ausgabe. In diesem Beitrag wird über die Neuentwicklung des PEARL-Compilers und die dabei benutzten Werkzeuge berichtet.

2. Das PEARL-Programmiersystem

Das portable Programmiersystem besteht aus dem PEARL-Compiler und den Laufzeitbibliotheken für das Tasking (BAPAS-K) und die Ein-/Ausgabe (BAPAS-EA). Von dem jeweiligen Zielsystem werden ein C-Compiler und der Systemlinker benutzt.

PEARL-Programme werden zunächst in die Programmiersprache C übersetzt, anschließend erfolgt eine Compilation mit einem C-Compiler, der einen linkfähigen Objektcode generiert. Durch den zum Zielsystem gehörenden Linker wird der Objektcode mit den PEARL- und C-Laufzeitbibliotheken zu einem ausführbaren Programm gebunden.

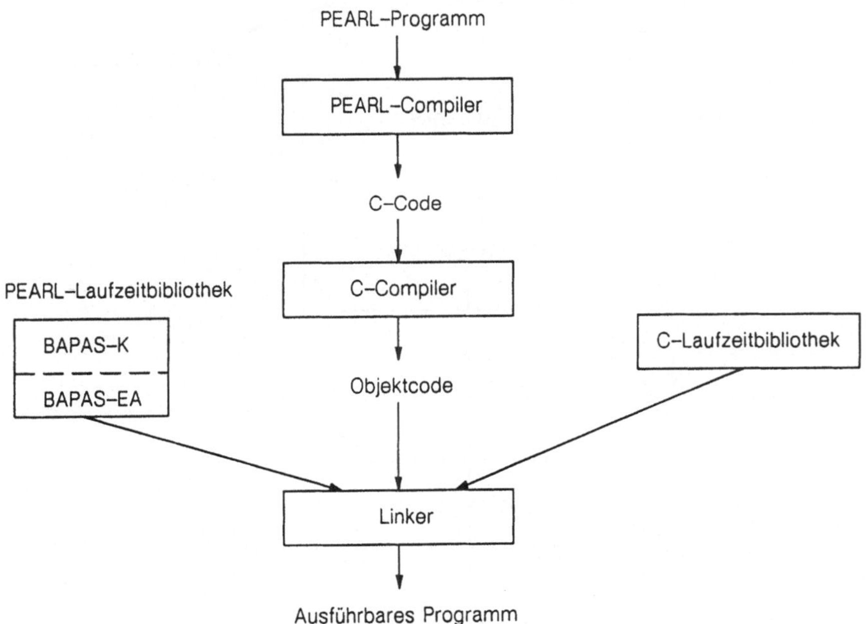

Bei einer Portierung auf ein neues Zielsystem müssen der PEARL-Compiler und das PEARL-Laufzeitsystem an das neue Betriebssytem angepaßt werden [Erdtmann 90]. Der

Compiler und die meisten Routinen der PEARL-Bibliothek liegen als portable PEARL-Programme vor, so daß nur einige wenige rechnerabhängige Funktionen jeweils neu implementiert werden müssen.

3.   Der neue PEARL-Compiler

Der neue Compiler ist als 3-Pass-Compiler konzipiert. In zwei Analysephasen werden Syntax und Semantik überprüft, ein dritter Pass generiert den gewünschten Zielcode. Die Syntax- und Semantikanalysephasen übersetzen ein PEARL-Modul in eine hochsprachennahe interne Darstellung. Der Codegenerator konvertiert diese interne Darstellung in eine gängige Hochsprache, für die auf der jeweiligen Zielmaschine ein Compiler vorhanden ist. Aufgrund der weiten Verbreitung von C-Compilern wurde bei dieser Implementierung die Sprache C als erste Zielsprache ausgewählt.

Das folgende Bild gibt einen Überblick über die einzelnen Compilerphasen, die mit "Syntax", "Semantik" und "Code-Generator" bezeichnet sind:

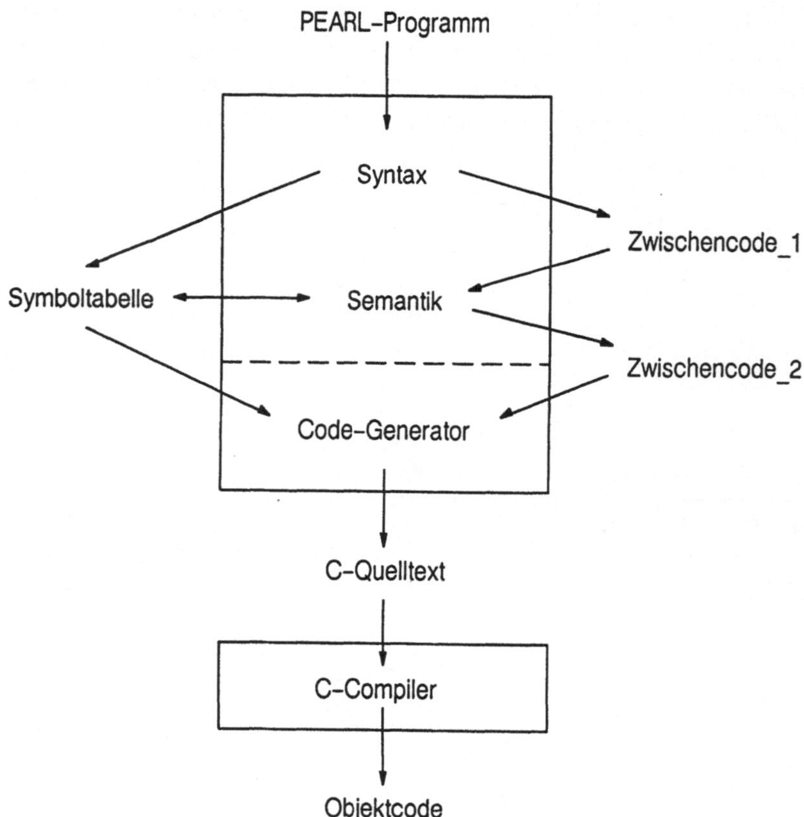

Die Wahl der Hochsprache C als Zielsprache für unseren Compiler hat folgende Gründe: Die Portierung auf ein neues Zielsystem erfordert wesentlich weniger Aufwand, als die Anpassung eines Assembler-Codegenerators. Dadurch sind wir in der Lage innerhalb kürzester Zeit unser PEARL-Programmiersystem auf ein neues Zielsystem zu portieren --- vorausgesetzt, es existiert bereits ein C-Compiler für die generierte Zielsprache. Ein weiterer, wichtiger Aspekt ist die schnelle Nutzung der neuesten Rechnerhardware. Da die Programmiersprache C bereits häufig zur Systemprogrammierung genutzt wird, existiert bei der Markteinführung neuer Rechner meist auch ein funktionsfähiger C-Compiler. So können wir kurze Zeit später ein PEARL-Programmiersystem für den neuen Rechner anbieten. Der verminderte Portierungsaufwand und die Vereinfachung der Wartung aller Implementierungen ermöglicht die Kalkulation eines günstigen Preises für das Gesamtsystem.

Auf den nächsten Seiten wird ein Überblick über die einzelnen Compilerphasen gegeben.

3.1   Die Syntaxanalyse

Der erste Pass führt die lexikalische Analyse und die Syntaxanalyse durch. Dabei werden auch Preprozessoranweisungen bearbeitet, die eine bedingte Compilierung und des Einfügen von Texten ermöglichen. Die Informationen über Blockstrukturen und über sämtliche Deklarationen werden in der Symboltabelle gesammelt. Die PEARL-Anweisungen und Ausdrücke werden in Listen und Baumstrukturen gespeichert. Die Anweisungen einer Prozedur bilden dabei jeweils die Elemente einer linearen Liste, in denen Verweise auf die enthaltenen Ausdrücke gespeichert sind. Die Operatoren und Operanden eines Ausdruckes werden in Syntaxbäumen gespeichert, die wir im weiteren als Expressionbäume bezeichnen.

3.2   Die Semantikanalyse

Im zweiten Pass werden die Semantikprüfungen durchgeführt. Bei der Überprüfung aller Deklarationen auf Vollständigkeit und Gültigkeit werden Symboltabellen-Einträge erweitert und vervollständigt. Die Statements werden blockweise ausgewertet und dabei die Semantik von Expressions überprüft (z.B. ob ein Operator für die aktuellen Operandentypen überhaupt definiert ist).

Die Expression-Bäume werden um implizite Operationen erweitert; dazu gehören z.B. Genauigkeitsverbesserungen von FIXED(15) auf FIXED(31) oder erlaubte Typwandlungen von FIXED nach FLOAT. Nach diesem zweiten Pass ist die eigentliche Compilation abgeschlossen, und das Programm liegt in einer hochsprachennahen internen Darstellung vor. Dabei werden in den Analysephasen (fast) keine Annahmen bezüglich der Zielsprache gemacht. Die Symboltabelle enthält nun Informationen über Blockstrukturen und Deklarationen, die verketteten Expressionbäume repräsentieren die Statements und Expressions.

3.3 Die Codegenerierung

Aus dieser hochsprachennahen internen Darstellung lassen sich leicht Abbildungen in die Kontrollstrukturen gängiger Hochsprachen gewinnen. Anstelle eines Hochsprachen-codegenerators könnte hier auch ein Pass zwischengeschaltet werden, der eine Lauf-zeitspeicherverwaltung vornimmt und so durch die Adressierung aller Objekte eine Abbildung auf eine Assemblersprache vorbereitet. Dieser Weg wurde nicht beschritten, da ein Assembler-Codegenerator für jedes neue Zielsystem umfangreiche Anpassungsarbeiten erfordert.

In der aktuellen Implementierung erzeugt der Codegenerator einen hochportablen C-Code und die Adressierung wird vollständig dem C-Compiler der Zielmaschine überlassen. C-Compiler sind für alle gängigen Rechner und Betriebsysteme vorhanden, und die Spache ist hinreichend genormt, so daß eine Portierung auf ein neues System keine Änderungen am Codegenerator erfordern wird. D.h. die Portierung des PEARL-Compilers besteht ausschließlich aus einer Neucompilierung der generierten C-Quellen auf dem neuen Ziel-system!

Die durch den C-Codegenerator erzeugten C-Quelltexte entsprechen weitgehend dem Standard nach Kernighan und Ritchie, den die meisten C-Compiler ohne Einschränkungen übersetzen können. Als einzige Erweiterung zu dem alten Standard wird vorausgesetzt, daß Komponenten mit gleichem Namen in verschiedenen Strukturen unterschieden werden. Da C-Compiler nach der ANSI-Norm noch nicht sehr verbreitet sind, werden diese Spracherweiterungen in der Standardeinstellung nicht generiert. Wahlweise kann jedoch ein ANSI-C konformer Code generiert werden (speziell Funktionsprototypen).

Außer der Sprache C sind aber auch Abbildungen auf andere Hochsprachen wie PASCAL oder Ada denkbar und mit vertretbarem Aufwand realisierbar. Dies ist sinnvoll z.B. für die automatische Umsetzung von existierenden PEARL-Programmen.

3.4 Anmerkungen zur Abbildung von PEARL auf C

Auf einige Unterschiede zwischen PEARL und C soll an dieser Stelle etwas genauer ein-gegangen werden. Die in Klammern erwähnten Bezeichner beziehen sich auf das weiter unten angeführte Programmbeispiel.

Die Sprache C erlaubt keine Prozedurdeklarationen innerhalb von anderen Prozeduren. Lokale Prozeduren (local_p) eines PEARL-Programmes müssen also bei der Generierung von C-Code außerhalb der aktuellen Prozedur (doit) deklariert werden. Dabei ergeben sich Probleme beim Zugriff auf Objekte, die innerhalb der umgebenden Prozedur deklariert wurden (bit16). Unsere Lösung für solche Fälle besteht in der Deklaration einer Struktur (_data), die alle lokalen Objekte einer Prozedur enthält. Beim Aufruf einer geschachtelten Prozedur wird ein Zeiger auf diese Struktur (_e1_doit) als ein zusätzlicher Parameter über-geben. Der Zugriff auf ein Objekt innerhalb der übergeordneten Prozedur geschieht dann über diesen Zeiger und den Namen des Objektes (_e1_doit->bit16).

Um keine Namenskonflikte mit anderen PEARL-Objekten zu bekommen und den Dokumentationswert zu erhöhen, beginnen die vom Compiler generierten Namen immer mit einem Unterstrich und einem Buchstaben zur Klassifizierung des Objektes (_t für Typen, _e für Environmentpointer, _p für lokale Prozeduren, usw.).

Das folgende Bild zeigt ein Beispiel für die Übersetzung von PEARL-Programmen nach C:

```
MODULE;                                              ┌───────┐
PROBLEM;                                             │ PEARL │
    DCL    bit    BIT(16);                           └───────┘

doit : PROC ( bit16 BIT(16) ) GLOBAL;
    DCL    maske   INV BIT(16) INIT ('0FF0'B4);

    local_p : PROC RETURNS ( BIT(16) );
        RETURN ( bit16 AND '00FF'B4 );
    END;

    bit := bit16 OR maske EXOR local_p;
END;

MODEND;



static   _BIT16  _p2_local_p ();                     ┌───────┐
static   _BIT16  bit ;                               │   C   │
                                                     └───────┘
typedef struct _t1__doit
  { _BIT16 bit16;
  } _t1__doit;

void  doit  ( bit16 )
    _BIT16  bit16;
{
  _t1__doit        _data;
  _data.bit16 := bit16;

    bit = (_data.bit16 | /*maske*/ 0x0FF0) ^ _p2_local_p (&_data);
} /*doit ()*/

static   _BIT16  _p2_local_p ( _e1_doit );
    _t1__doit * _e1_doit ;
{
  return _e1_doit->bit16 & 0x00FF;
} /*_p2_local_p ()*/
```

Des weiteren zu beachten sind die unterschiedlichen Operatorprioritäten in C und PEARL. Beim Aufbau der Expressionbäume in den Analysephasen werden die PEARL-Prioritäten und eventuelle Klammerungen von Subausdrücken berücksichtigt. Die Struktur des erzeugten Expressionbaumes spiegelt die Auswertungsreihenfolge der Operatoren wieder, Klammern sind nicht mehr enthalten. Erst bei der Generierung von C-Code aus einem Expressionbaum werden die Prioritätsklassen von C und deren implizite Klammerungsreihenfolge berücksichtigt. Hierbei werden evtl. notwendige Klammerungen eingefügt. Anhand eines Beispiels soll dieser Fall dargestellt werden:

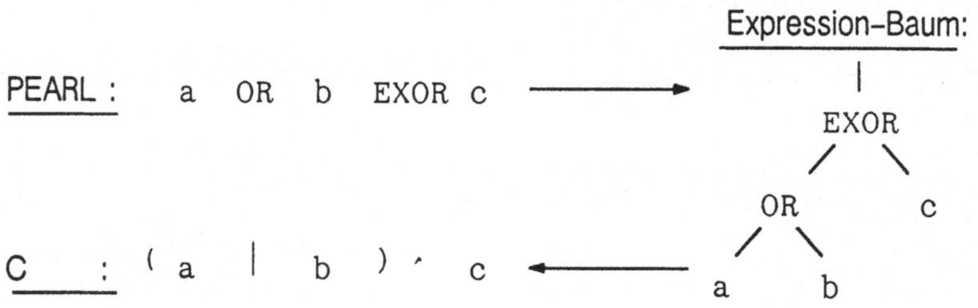

Expression-Baum:

```
                                              |
PEARL :    a  OR  b  EXOR c  ———————————→   EXOR
                                            /    \
                                         OR        c
                                        /  \
C    :  ( a  |  b  )  ^  c  ←—————————— a    b
```

In C hat der EXOR-Operator ("^") höhere Priorität als der OR-Operator ("|"), deshalb muß bei der Abbildung des Expressionbaumes in die Sprache C der erste Teilausdruck geklammert werden.

4.   Hilfsmittel

Als Hilfsmittel wurden bewußt keine der gängigen Compilergeneratoren eingesetzt. Speziell aus der UNIX-Welt dürften die Programme lex und yacc (Scanner- und Parser-Generator) bekannt sein, die Analyseprogramme in der Sprache C erzeugen. Die gewünschte hohe Portabilität läßt sich aber nur erreichen, wenn auch die generierten Programmteile überall einsetzbar sind. Deshalb kamen von vornherein nur Werkzeuge in Betracht, die direkt PEARL-Programme oder Teile eines PEARL-Programmes (z.B. Initialisierungen von Feldern) generieren können. So ist die gewählte Zielsprache C nur eine von vielen möglichen Alternativen und nicht durch die benutzten Werkzeuge vorgegeben.

4.1   Die Beschreibung der PEARL-Syntax

Die Syntaxanalyse läuft in Form einer tabellengesteuerten LL(k)-Analyse mit Backtracking ab. Die benutzte Steuertabelle wird durch ein Generatorprogramm aus einer in lesbarer Form vorliegenden Grammatik erzeugt. Diese Grammatik beschreibt in einer Backus-Naur-Form die PEARL-Syntaxregeln, sowie Regeln für die Syntax der Preprozessorstatements. Es sind sogar Regeln für die Auswertung der Kommandozeile beim Compilerstart enthalten.

Das Generatorprogramm --- eine Weiterentwicklung eines Tools, das im Hause Werum bereits seit mehreren Jahren genutzt wird --- erstellt neben der Steuertabelle weitere Dateien: z.B. Konstantendeklarationen für Kodierungen von Schlüsselworten und andere lexikalische Elemente, sowie Funktionsaufrufe von Semantikfunktionen.

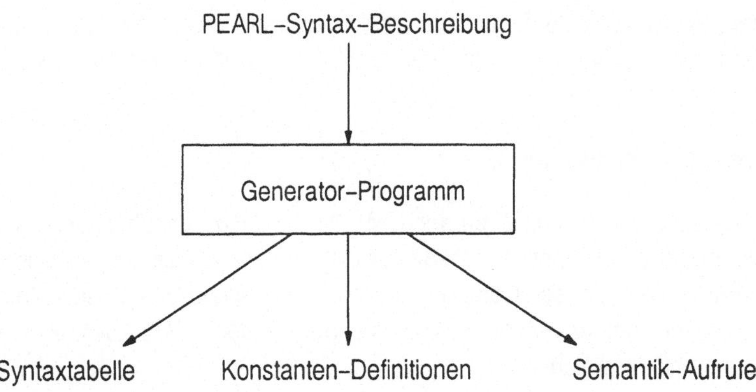

Änderungen der Syntaxbeschreibung führen nach Einbringung der Steuertabelle in den neuen Compiler (Recompilierung eines Moduls) automatisch zu einem Syntaxanalyseprogramm, das die nun veränderte Sprache akzeptiert. So sind Änderungen oder Erweiterungen der Sprache sehr komfortabel möglich.

```
/* ---------- PROBLEM DIVISION ---------------------------------- */


problem_division:
        PROBLEM SEMI               &sy_in_declaration_part
                                   &sy_st_enter_main_block
        ( declaration  SEMI )*     &sy_st_leave_main_block
        ;


/* ---------- DECLARATIONS -------------------------------------- */


declaration:
        length_definition
      | declare_sentence
      | label  (  procedure_declaration
               |  task_declaration
               )
      | type_definition
      | specification
        ;
```

Das obige Bild zeigt einen Ausschnitt aus der Syntax-Beschreibungsdatei. Die beiden Worte PROBLEM und SEMI (";") bezeichnen zwei Terminalsymbole von PEARL, die kleingeschriebenen Identifier sind jeweils die Namen von Syntaxregeln. Mit Klammern kann man mehrere Elemente zusammenfassen, und ein folgender Stern ("*") zeigt an, daß das Element oder die geklammerte Gruppe beliebig oft (auch nullmal) vorkommen darf. Ein senkrechter Strich trennt die Alternativen einer Regel. Die auszuführenden Semantikfunktionen, die während der Syntaxanalyse z.B. Eintragungen in die Symboltabelle vornehmen, werden in der Beschreibungsdatei durch ein vorangestelltes "&" gekennzeichnet.

4.2 Der LL(1)-Analysator

Mit einem Analyseprogramm überprüfen wir die per Hand geschriebene Grammatik auf LL(1)-Verletzungen. Bei einer LL(1)-Syntaxanalyse dürfen insbesondere keine Linksrekursionen vorkommen, da der Compiler sonst in eine unendliche Schleife geraten kann. Weiterhin bezeichnet das Analyse-Tool die Stellen der PEARL-Syntax, die von einer LL(1)-Grammatik abweichen.

Ein Beispiel für die Verletzung des LL(1) Kriteriums liefert folgender Ausschnitt aus der PEARL-Syntax:

```
block       :  BEGIN  (declaration)*  (statement)*  END

declaration:  ...  |  ID ':' PROC ...  |  ...

statement   :   ( ID ':' )*  unlabelled_statement
```

Das Terminalsymbol "ID" kann sowohl der Beginn einer Prozedurdeklaration, als auch der Beginn eines mit einem Label versehenen Statements sein. Das folgende Symbol, der Doppelpunkt (":"), gibt auch noch keinen Aufschluß über die vorliegende Syntaxstruktur. Erst mit dem dritten Symbol kann man diese Fälle unterscheiden: liegt das Schlüsselwort "PROC" vor, so ist es eine Prozedurdeklaration, sonst der Beginn eines mit einem Label versehenen Statements.

Nach der Analyse der von einer LL(1)-Grammatik abweichenden Syntaxstrukturen kann man evtl. einen Teil der Regeln in eine solche Grammatik verbessern. Anderenfalls kann man abschätzen, wie häufig ein Backtracking in der Analyse vorkommen wird. Der oben angeführte Fall tritt relativ selten auf, nämlich bei der Übersetzung eines PEARL-Programmes pro Block genau einmal, und ist daher vernachlässigbar.

4.3 Der Memory-Manager

Erfahrungen aus anderen Compiler-Projekten haben gezeigt, daß die Speicherverwaltung für die internen Tabellen möglichst flexibel realisiert werden sollte. In den Compiler ist ein Speicherverwaltungstool integriert, das (von PEARL aus) dynamische Speicheranforderungen ermöglicht. Dieses Programm verwaltet sowohl den verfügbaren Hauptspeicher, als auch einen Hintergrundspeicher in Form einer Plattendatei mit Direktzugriff.

Dabei wird der angeforderte Speicher in Größen von etwa 1 bis 8 KB (Units) aufgeteilt. Bei einer Anforderung vom Compiler wird ein freier Bereich zur Verfügung gestellt, falls nötig wird vom Betriebssystem Speicher verlangt (malloc-Funktion), oder es wird ein z.Z. nicht benötigter Speicherbereich in die Hintergrunddatei ausgelagert und der nun freie Bereich erneut verwendet. Erfolgt ein Zugriff auf ein ausgelagertes Unit, so wird dieses durch den Memory-Manager automatisch wieder eingelesen; evtl. wird dabei ein anderer Bereich in den Hintergrund verdrängt.

Dieses Tool soll den Compiler unabhängig von dem zur Verfügung stehenden (Haupt-) Speicher machen. Insbesondere ermöglicht es dem Compiler, größere Programme auch auf Rechnern mit wenig Hauptspeicher, bzw. ohne virtuelle Speicherverwaltung, zu übersetzen.

5. Ausblick

Bei der Neuentwicklung des Compilers wurde großer Wert auf die Flexibilität und Wartbarkeit der Programme gelegt. Durch die Verwendung von lesbaren Syntaxbeschreibungen und automatischer Generierung der Steuertabelle für die Syntaxanalyse lassen sich Änderungen des Sprachumfanges sehr leicht in den Compiler einarbeiten.

Zunächst wurde der Sprachumfang des alten Compilers mit einigen PEARL 90-Erweiterungen implementiert. Die Erweiterungen auf den vollständigen Sprachumfang von PEARL 90 werden folgen, sobald eine genaue Festlegung von PEARL 90 und Beschreibungen der Syntax und Semantik existieren.

Als erste Zielsysteme sind IBM PS/2 unter OS/2, VAX unter VMS, sowie diverse UNIX-Systeme vorgesehen.

Literatur

[Erdtmann 90]
  Erdtmann, A., Kneuer, E.: Implementierung der Echtzeitsprache PEARL unter dem Echtzeit-UNIX-Betriebsystem SORIX von Siemens. In: Drebinger, L. (Hrsg.): Echtzeit '90, Kongreß-Vortrags-Band, 19. bis 21. Juni 1990, Sindelfingen.

[Kneuer 88]
  Kneuer, E.: Erfahrungen mit der Portierung eines Prozeßleitsystems. Tagungsband zur PEARL-Tagung '88, 1. und 2.12.1988, S. 109 – 118. Hrsg.: PEARL-Verein.

[Stieger 89]
  Stieger, K.: PEARL 90 – Die Weiterentwicklung von PEARL. In: Informatik Fachberichte 231, PEARL '89 Workshop über Realzeitsysteme, Springer Verlag, 1989.

[Werum 89]
  Werum, W.; Windauer, H.: Introduction to PEARL. Vieweg 1989, 4. Auflage.

Leistungsmerkmale des Realzeit-UNIX-Systemes R E A L / I X

Peter Guba

Modular Computer Systems GmbH
Geschäftsstelle Konstanz
Bücklestraße 1-5
7750 Konstanz
Telefon: (07531) 807-449

<u>Zusammenfassung:</u>

REAL/IX ist konform zur UNIX System V Interface Definition (SVID)
und wurde gemäß System V Verification Suite (SVVS) geprüft.

REAL/IX übertrifft funktional den Industriestandard UNIX V.3 .

Zahlreiche Erweiterungen sind auf die speziellen Anforderungen für
Realzeitanwendungen ausgerichtet.

Um aus dem UNIX V.3 von AT&T ein Realzeitsystem zu machen, wurden folgende Erweiterungen integriert:

- Voll unterbrechbarer Betriebssytem-Kern
- Echtzeit- und Timesharing Prioritäten in einem Betriebssystem
    . feste Prioritätsebenen für die Realzeitprozesse
    . "herkömmliche" Prioritäten für die Timesharing-Prozesse
- erweiterte Memory-Management-Funktionen
- erweiterte und optimierte Interprozeßkommunikation
- erweiterte Realzeit-Timer-Mechanismen
- erweiterte "Signal"-Mechanismen
- erweitertes File-System
    . zusätzliches schnelles File-System
    . Möglichkeit, den Pufferspeicher zu umgehen
    . Möglichkeit, zusammenhängende Plattenbereiche zu beschreiben
    . größere logische Blocklängen
- Asynchrone Ein-Ausgabeverwaltung

Das Betriebssystem ist ablauffähig auf den Computersystemen MC 97xx der
Firma Modular Computer Systems GmbH. Es ist Binärcode-kompatibel zu
UNIX V/68 von Motorola.

Literatur:
Concepts and Characteristics - REAL/IX Operating System
MODCOMP publication 205-855001-000, 1989

Die Kosten der Software übersteigen bei den heutigen Realzeitsystemen fast immer die Kosten der Hardware. Diese Tatsache zwingt auch bei der Systemerstellung von Realzeitsystemen zur Standardisierung.

Der Weg dahin wird durch die Nutzung eines herstellerunabhängigen Betriebssystems (UNIX), inzwischen anerkannter gehobener Programmiersprachen (Ada, C, F77) sowie die Nutzung von standardisierter Hardware (Motorola 68000) erleichtert.

Offene Systeme reduzieren die Systemkosten und die notwendige Entwicklungszeit, da es möglich ist auf vorhandene Softwarepakete mit einfacher Integration und Software-Portabilität zurückzugreifen.

Das Betriebssystem UNIX ist inzwischen ein Standard geworden, der durch seine hohe Flexibilität und die große Anzahl zur Verfügung stehender Tools eine sehr hohe Akzeptanz erreicht hat. Zusätzlich bietet UNIX den Herstellern von Software eine hervorragende Entwicklungs- und Programmierumgebung.

Obwohl das Standard-UNIX-Betriebssystem für Multitasking und Timesharing konzipiert ist, ist es nicht in der Lage, für zeitkritische Applikationen einen großen I/O- Durchsatz sowie eine garantierte Antwortzeit zu gewährleisten.

Diese Eigenschaften wurden in der Implementation des Betriebssystems REAL/IX von MODCOMP integriert.

Der Einsatz dieser Standards erleichtert die Portierung der Software bei der Umstellung auf leistungsfähigere Hardware und reduziert die Ausbildungskosten der Systemanalytiker (Programmierer).

Dieser Industriestandard wird die Basis für die nächste Generation der Realzeitsysteme bilden.

Vorteile offener Systeme gegenüber herstellergebunder Systeme

Vorteile für den Nutzer	Herstellergebundene Systeme	Offene Systeme
Übertragungsdauer von Software	Monate / Jahre	Stunden / Wochen
Datenbank Anpassungen	Jahre	Stunden / Tage
Ausbildung der Programmierer	häufige Ausbildungszeiten (Umschulungen)	zu vernachlässigen
Leistungssteigerung	Abhängigkeit vom Hersteller	Freier Markt mit großer Innovation

<u>Welche Vorteile hatten herstellerbezogene Realzeit-Betriebssysteme:</u>

- Die Programme wurden in der Realzeitumgebung entwickelt

- Die Systeme hatten ein garantiertes Zeitverhalten auf Ereignisse

- Die Systeme waren für die Realzeitanwendungen "getuned"

- Die Systeme zeigten spezielle Realzeiteigenschaften.

<u>Welche Vorteile hatten reine UNIX - Systeme:</u>

- Hohe Standardisierung des Betriebssystems

- Hohe Portabilität der Software

- Ein großes Angebot an Tools und Applikationen

- Hohe Akzeptanz bei den Programmierern

<u>Wie wurde in der Vergangenheit UNIX realzeitfähig?</u>

- Durch Kopplungen UNIX-Entwicklungssystem mit Realzeit-Zielsystem

- Durch herstellereigene UNIX Implementierungen

- Durch Erweiterungen von UNIX Richtung Realzeit

- Durch Unterbrechbarkeitspunkte (Fenster) im Betriebssystemkern

- Durch volle Unterbrechbarkeit des Kernes mit Integrität der
  Datenstrukturen

R E A L / I X   MODCOMP's Realzeit - UNIX - Betriebssystem
==

<u>Das Betriebssystem REAL/IX bietet folgende Vorteile :</u>

- AT&T UNIX System V Release 3

- Erweiterung um Realzeit-Eigenschaften und -Leistungen

- Integrierte Programmierumgebung auf dem Zielsystem

- Einzel- und Mehrprozessorfähigkeit

- Ausrichtung auf den Standard von POSIX 1003.4

- Binärkompatibel zum System V/68 von Motorola

- Ausgewogene Tri-Dimensional Leistung für MIPS, I/O-Operationen und
  Interrupthandling

Im folgenden wird der Weg vom Standard UNIX V/3 zu dem Hochleistungs-
realzeitsystem REAL/IX aufgezeigt.

1. Erweiterung der Prozeßverwaltung

Die Erweiterung der Prozeßverwaltung umfaßt einen vollunterbrechbaren
Betriebssystemkern, eine prioritätsgesteuerte Prozeßverwaltung, garantierte Interruptzeiten, realzeitorientierte TimerMechanismen sowie ein
erweitertes Memory-Management.

-- Unterbrechbarer Betriebssystemkern

Ein unterbrechbarer Betriebssystemkern garantiert die sofortige Bearbeitung eines Prozesses oder Interrupts hoher Priorität, und zwar
unabhängig davon, ob sich die CPU im Anwender- oder Betriebssystem-
Modus befindet.

Im Gegensatz dazu wird bei Standard-UNIX erst ein zur Zeit ausgeführter
Kernel-Aufruf komplett durchgeführt, bevor die höherpriore Task fortgesetzt oder begonnen wird.

-- Prioritätsgesteuerte Prozeßverwaltung

Ein prioritätsorientiertes Scheduling-System erlaubt es dem Benutzer,
die Prioritäten der einzelnen Prozesse festzulegen. Hierdurch erfolgt
die Prozeßteuerung bei Interrupts je nach der vom Benutzer eingestellten Priorität.

Im Gegensatz dazu wird bei den meisten Multitasking-Systemen die
Priorität vom System selbst vergeben.

Aus diesem Grund haben Realzeitprozesse bei REAL/IX eine feste
Priorität im Bereich von 1 bis 126, Timesharing-Prozesse eine freie
Priorität im Bereich von 128 bis 254.

-- Garantierte Interruptzeiten

Das Realzeitsystem muß in der Lage sein, ein anstehendes Ereignis
möglichst schnell zu erkennen und dann die notwendige Aktion
auszuführen.

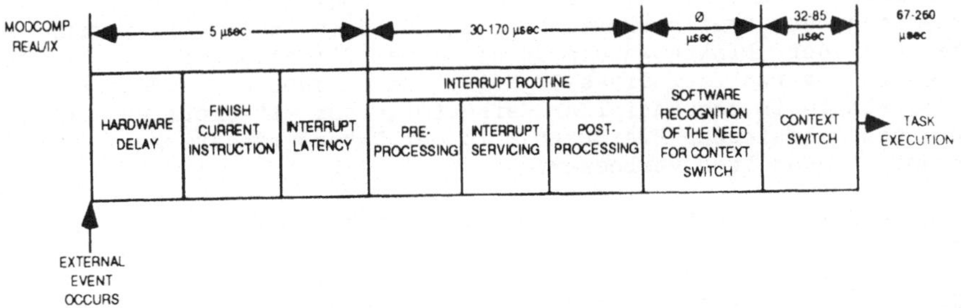

Das System muß auf Hardware und Software-Interrupts reagieren können.
Das System muß sich selbst unterbrechen können. Da darf nur ein mini-
maler Overhaed im Betriebssystem sein, besonders bei einem Contex-
Switch zu einem hochprioren Realzeitprozeß, der ausführbar ist.

-- realzeitorientierte Timer-Mechanismen

Die Realzeittimer bieten Realzeitprozessen die Möglichkeit, sich über
System-Events in sehr kurzen Zeitabständen (von einigen Sekunden bis zu
12,5 µs) wecken zu lassen. Diese Timer sind absolut oder relativ
möglich. Außerdem können diese Timer als einmaliger Auftrag oder als
periodischer Auftrag ausgeführt werden.

Zusätzlich steht der Standard-UNIX-Timer (Sekunden und Nanosekunden
seit 1.1.1970 ebenfalls zur Verfügung).

-- eine erweitertes Memory-Management.

Vorbelegung von virtuellem Speicherplatz
Die Speicherbelegung entspricht den Funktionen bei Standard UNIX.
Zusätzlich steht für zeitkritische Realzeitprozesse die Möglichkeit der
Kontrolle über den Speicher zur Verfügung. Um eine garantierte Responz-
zeit zu erreichen, ist es dem Benutzer möglich, eine Vorbelegung
(preallocate) von virtuellem Speicher für den Datenteil zu erreichen.
Viele UNIX-Systeme erlauben schon die Vorbelegung von virtuellem Speicherplatz für den Datenteil. REAL/IX erlaubt zusätzlich die Vorabzuwei-
sung für den Stack.

Reservierung von Speicher
Für zeitkritische Prozesse steht ein Systemaufruf zur Verfügung, der
das Paging von Speicherteilen (Text, Daten, Shared-Memory) ausschaltet.
Hierbei werden die Teile im Speicher gehalten, nachdem sie das erste
Mal benötigt wurden. Sie können aber auch mit dem Aufruf für die Spei-
chervorbelegung direkt in den Speicher geladen werden.

2. Erweiterung des File-Systems

Das Betriebssystem REAL/IX bietet dem Anwender zwei Dateisysteme: das
UNIX - Standarddateisystem und ein realzeitfähiges schnelles Dateisystem.

Das schnelle Dateisystem bietet die Möglichkeit, physikalisch
zusammenhängende Speicherbereiche für Dateien zu reservieren. Die
Blöcke auf der Platte werden durch eine Bit-organisierte Platten-
organisation verwaltet. Diese Organisation erlaubt es, auch die Blöcke
auf der Platte in sequentieller Reihenfolge und mit größeren logischen
Blocklängen abzulegen. Hierdurch wird der Datendurchsatz bei größeren
Datenmengen wesentlich verbessert.

Der in UNIX realisierte Pufferspeicher kann bei der Daten-E/A umgangen
werden.

Das System kennt sowohl synchrone als auch asynchrone Dateiaufrufe. Bei
der asynchronen Datei-E/A wird der aufrufende Prozess mit dem "common
even mechanism" fortgesetzt, oder er pollt den Kontrollblock des Auf-
rufes.

Im Standard-UNIX werden die Ein/Ausgaben auf die Platte im Hauptspeicher (Puffer- oder Cache-Speicher genannt) zwischengepuffert
(hierher umkopiert). Der Datentransfer erfolgt durch ein zyklisches
Programm oder wenn die Organisation des Speichers "neuen" Platz
benötigt. Diese Organisation ermöglicht es, bei kleinen Blocklängen und
Schreib- und Leseaufrufen auf die gleichen Blöcke ohne Datentransfer
auszukommen.

Bei großen Datenmengen bewirkt diese Organisation das überflüssige Um-
speichern von großen Datenblöcken.

3. Erweiterung des E/A-Systems

Das Betriebssystem REAL/IX unterstützt asynchrone E/A-Operationen.
Hierdurch lassen sich Überlappungen in der Prozeßbearbeitung und der
Ein/Ausgabe erreichen. Mit dieser Möglichkeit können die Ausführungs-
zeiten von Prozessen wesentlich verkürzt werden.

Durch die Möglichkeit von prioritätsgesteuerten Warteschlangen bei
Plattenzugriffen lassen sich höherwertige Tasks - gegenüber solchen mit
geringerer Priorität - bevorzugt bedienen.

Eine direkte Ein/Ausgabe ermöglicht es Anwenderprogrammen, E/A-Daten im
eigenen Adreßraum abzubilden. Dieser Vorteil des Systems senkt die Belastung durch die E/A-Operationen sobald diese Bereiche im Anwenderprogramm plaziert sind, da Umspeichervorgänge entfallen.

Die Nutzung des "connected interrupt" ermöglicht es, einen Device-
Driver als Anwenderprozeß zu realisieren. Diese Eigenschaft der Benut-
zerebene ist sehr nützlich, wenn eine enge Kontrolle mit dem Device
notwendig ist.

Die Verbindung zwischen dem Betriebssystem (nach einem Interrupt) und
dem Anwenderprogramm erfolgt über Datenstrukturen oder über den "common
event mechanism".

4. Erweiterte Interprozeß-Kommunikation

-- "common event notification mechanism"

Der "common event notification mechanism" ist ein erweiterter Signal-
Mechanismus, mit dem Botschaften zwischengespeichert werden können.
Diese Botschaften können synchron als auch asynchron empfangen werden.

Events können vom Prozeß selbst, von anderen User-Prozessen oder vom
Kernel aus geschickt werden. Eine 32-Bit Dateninformation kann mit jedem Event mitgegeben werden. Jeder Event-Warteschlange wird ein eindeu-
tiger Event-Identifier zugeordnet. Wie bei Signals ist diese Event-
Queue nur dem empfangenden Prozeß zugeordnet. Da sie zwischengespeichert werden, gehen Events an di e gleiche Event-Queue nicht verloren.

-- "binary semaphore mechanism"

Der größte Teil aller Semaphorenoperationen erfolgt nur zur Koordination gemeinsamer Bereiche zwischen mehreren Prozessen. Die Operationen sind in 95 % dieser Fälle überflüssig, da sie nur in
Konfliktfällen benötigt werden. Bei dem "binary semaphore mechanism"

wird dieser Punkt berücksichtigt. Im Normalfall führt die Belegung eines gemeinsamen Bereiches nur zum Setzen einer Markierung (direkt aus der eingebundenen Library des Anwenderprogrammes heraus - ohne Aufruf des Betriebssystems). Nur bei einem Konfliktfall wird ein Semaphoraufruf ausgeführt.

Unabhängig von diesen "Beleg-Optimierungen" ist der Mechanismus der "binary semaphore" für die anderen Semaphoranwendungen wesentlich schneller als der Standard-UNIX-Aufruf.

5. Bourne - und Korn - Shell

Die gebräuchlichsten Schnittstellen zum Betriebssystem ist die Shell, ein Kommandointerpreter, der die Eingaben des Benutzers verarbeitet.

Hierbei bietet die Shell eine leistungsfähige Programmiersprache mit Variablen, Unterroutinen und Parameterübergaben.

Das Betriebssystem REAL/IX bietet zwei Shells an:

-- Bourne Shell

Die Bourne Shell ist die Standards Shell des Systems UNIX System V. Sie ist anwenderfreundlich und flexibel beim Programmieren.

-- Korn Shell

Die Korn Shell kombiniert die Vorteile der Bourne Shell und der C-Schell (BSD Systeme) mit etlichen wichtigen Erweiterungen. Sie ist Bestandteil des Betriebssystems REAL/IX und dürfte künftig zum Standard werden.

6. TCP/IP Protokoll

Das Betriebssystem REAL/IX unterstützt das TCP/IP-Protokoll und die Utilities zur Realisierung von Aufgaben in lokalen Netzwerken. Es enthält die telnet und ftp - Kommandos und die Berkeley remote-Kommandos. Das Betriebssystem stellt eine Berkely-Socket-Schnittstelle für die Programmierung der TCP/IP-Funktionen zur Verfügung.

7. X Window System

Das Betriebssystem REAL/IX unterstützt das netzwerkbasierende Grafik-Window-System X-Windows. Das System stellt die Client-Seite der X-Window-Routinen zur Verfügung, während der X-Server die serverseitigen Routinen zur Verfügung stellt. X-Server können X-Window-Terminals oder PC's sein, die über TCP/IP und Ethernet verbunden sind.

Mit dem X-Window-System unterstützt das Betriebssystem REAL/IX das Motif-Paket.

Das Graphik-Konzept des SIMICRO SX
- Die erste Entwicklung eines "Prozeß-X Windows"

Wolfgang Trennhaus
Siemens AG
Bereich Automatisierungstechnik
Geschäftsgebiet Steuerungs- und Prozeßregelsysteme
Microcomputersysteme
Siemens AG AUT V151
Postfach 4848
D-8500 Nürnberg 1

Zusammenfassung:

Beim Bereich Automatisierungstechnik der Siemens AG wurde für die
Anwendung in der Prozeßautomatisierung der SIMICRO SX entwickelt.
Einer der wichtigsten Grundsätze hierbei war es, eine Verwendung
von allgemein akzeptierten Standards mit den für die Prozeßautomatisierung notwendigen Erweiterungen zu verbinden. Im Bereich
der Prozeßautomatisierung kommt der graphischen Prozeßvisualisierung immer größere Bedeutung zu, deshalb wurde auch für
den SIMICRO SX eine Möglichkeit geschaffen, graphische Darstellungen von Prozessen zu gestatten. Für UNIX-Systeme hat sich
das X Window System als allgemein akzeptierter Standard für
Graphik-Funktionen etabliert. Es liegt also nahe, eine um Echt-
zeit-Eigenschaften erweiterte Version des X Window Systems als
Grundlage für die Graphik des SIMICRO SX zu entwickeln. Im folgenden Aufsatz wird zunächst kurz auf Historie und grundsätzliche
Eigenschaften des X Window Systems eingegangen. Anschließend
werden anhand von Systemeigenschaften von X Windows die notwendigen Erweiterungen für die Prozeßvisualisierung hergeleitet.
Weiterhin wird die Hardware-Basis der Graphik des SIMICRO SX behandelt. Schließlich wird das Software-Paket XMOVE vorgestellt,
das für den SIMICRO SX entwickelt wurde und die Erstellung von
Prozeßbildern aufsetzend auf dem X Window System durch einen
objekt-orientierten Graphik-Editor unterstützt.

Einleitung

Im Bereich der Prozeßautomatisierung, der Domäne des SIMICRO SX,
nimmt die Möglichkeit, mit Hilfe von Graphik-Terminals Prozeßvisualisierung zu realisieren immer höheren Stellenwert ein. Der
Mensch, und somit auch Personen, die Prozesse mit Hilfe von Micro-
computern steuern, ist von seiner Anlage her visuell orientiert.
Er kann Prozeßzustände visuell erheblich schneller überblicken und
analysieren, wenn sie in graphisch aufbereiteter Form dargestellt
sind. Auch hier hat der alte Ausspruch Gültigkeit: "Ein Bild sagt
mehr als tausend Worte." Diesem Trend hin zur graphischen Prozeßvisualisierung trägt auch der SIMICRO SX Rechnung, und stellt
Mittel zur Visualisierung von Prozessen über das X Window System
(im folgenden kurz X genannt) zur Verfügung.

Das X Window System - Entstehung und grundsätzliche Funktionen

Das X Window System entstand am MIT[1] in den frühen achtziger
Jahren aus einer Kooperation zwischen verschiedenen Rechner-
herstellern (DEC, HP,...) und dem MIT, hat sich jedoch im Laufe
der Jahre zu einem allgemeinen, herstellerunabhängigen Standard
entwickelt. Dem trägt Rechnung, daß Erweiterungen und zusätzliche
Funktionen von einem Konsortium, dem sog. X-Konsortium, welches
sowohl Firmen als auch Universitäten offensteht, definiert werden.
Die Graphik-Schnittstelle X Windows fand auch Eingang in das von
X/OPEN normierte Common Application Environment CAE. Derzeit
aktuell ist die Version 11.4 dieses Systems, wobei der Löwenanteil
der am Markt verfügbaren Produkte noch auf der Version 11.3 auf-
setzt.
Welche Funktionalität bietet nun dieses X Window System seinen Anwendern? X Windows bietet ein verteiltes Konzept für die Realisierung von Graphik. Verteilt bedeutet hier, daß von Anwendungen, die
auf verschiedenen Rechnern am LAN ablaufen, Ausgaben auf Graphik
Terminals erfolgen können. Wie wird dies erreicht? Grundsätzlich
unterscheidet X zwei Instanzen, die Clients und den X Server.
Clients geben Graphik- und Window-Aufträge über eine Library-
Schnittstelle, die X Library, an den X Server ab, der diese dann
am Graphik-Bildschirm darstellt. Die X Library verwendet zur Kom-

---

[1] Massachussets Institute of Technology

munikation zum Server ein aus dem Berkeley UNIX stammendes Mittel, die Sockets. Diese Sockets können sowohl auf dem gleichen Rechner ablaufende Prozesse adressieren, als auch über TCP/IP über ein LAN auf einem anderen Rechner ablaufende Prozesse erreichen. Hierdurch ergibt sich die erfreuliche Eigenschaft, daß eine Anwendung nicht wissen muß, ob der X Server auf dem gleichen Rechner abläuft oder auf einem anderen Rechner, die Ortstransparenz für X-Applikationen. Wichtig ist in diesem Zusammenhang, daß über die Socket-Schnittstelle Aufträge und Daten in komprimierter Form über eine spezielle normierte Schnittstelle, das X-Protokoll, transportiert werden, also die Belastung durch Kommunikation minimiert wird. Zusätzlich zur X Library gibt es noch eine Anzahl von diese benutzenden Zusatzbibliotheken, die Toolkits, von denen ein bekannter Vertreter OSF/Motif, das von der Open Software Foundation (OSF) definierte Standard-Toolkit ist. Eine ganze Reihe von Standard-Applikationen rundet das vorhandene Software-Spektrum auf der Client-Seite ab. Zwei dieser Clients möchte ich exemplarisch näher erläutern, weil sie besonders wichtig für das X-Window System sind. Als erstes möchte ich auf das sog. Xterm, eine Character-Terminal-Emulation für das X-Window System, eingehen. Xterm erlaubt es, auch für Anwendungen, die nicht speziell für die Graphik-Schnittstelle geschrieben wurden, auf einem Graphik-Bildschirm Ausgaben zu ermöglichen. Dies wird dadurch erreicht, daß dieses Programm sich der Anwendung gegenüber wie ein normales Character-Terminal verhält, Ausgaben jedoch in ein Window auf dem Grafikbildschirm weiterleitet. Eine weitere Applikation von zentraler Bedeutung für das X Window System ist der sogenannte Window-Manager. Ein Window-Manager ist dasjenige Programm, das den Bildschirm für die clients verwaltet. Anforderungen zum Vergrößern und Verschieben von Fenstern werden vom System an den Window-Manager weitergeleitet, der sie dann gegebenenfalls ausführt. Im Standard-Umfang der X Window Pakete sind verschiedene Window-Manager mit unterschiedlichen Eigenschaften enthalten; verschiedene Anwendungsbereiche erfordern hier eventuell auch unterschiedliche Windowmanager.

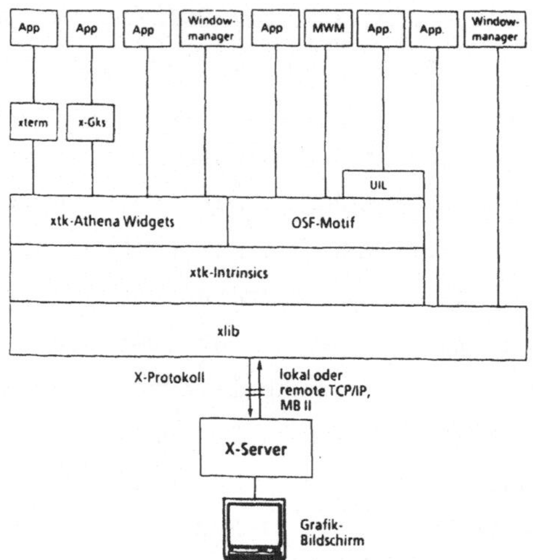

Bild 1 : Der Aufbau des X Window Systems

Erweiterungen des X Window Systems für die Prozeßvisualisierung

Für Anwendungen in der Prozeßautomatisierung entstand bei Siemens der SIMICRO SX mit dem von Siemens entwickelten Echtzeit-UNIX[2]- Betriebssystem SORIX[3]. Aus dem spezifischen Anwendungsprofil für die Automatisierung ergeben sich auch spezifische Anforderungen an die Graphik. Dies soll jedoch nicht darüber hinwegtäuschen, daß sich weite Teile der Anforderungen mit denen anderer Anwendungen decken. Es ist also wünschenswert, aufsetzend auf einen Standard die Zusatzanforderungen als Erweiterungen zu realisieren. Im folgenden möchte ich einige wichtige Anforderungen der Prozeß- automatisierung formulieren und am Standard X Window spiegeln. Hierdurch ergeben sich die Ansätze für Erweiterung des X Window Konzeptes, die für SORIX geplant bzw. realisiert sind.
Als Hauptanforderung der Prozeßautomatisierung kann man Echt- zeitfähigkeit und Verläßlichkeit ansehen. Echtzeitfähigkeit be- deutet garantierte Reaktionszeiten auf äußere Ereignisse, also für die Prozeßvisualisierung, daß der Anwender berechnen, bzw. ab- schätzen kann, daß und wann eine Ausgabe am Bildschirm erscheint. Hierbei geht es jedoch weniger um Reaktionszeiten im mikro- oder millisekunden- Bereich als um Zeiträume im Sekundenbereich, denn ein menschlicher Prozeßbeobachter nimmt kürzere Zeiten als 300

---

[2] UNIX ist ein eingetragenes Warenzeichen von AT&T
[3] SORIX ist ein eingetragenes Warenzeichen von Siemens

millisekunden ohnehin nicht wahr und egalisiert durch seine eigene
Reaktionszeit ein solches Zeitverhältnis ohnehin. Des weiteren
sollte man einem menschlichen Beobachter nur die Steuerung von
Prozessen aufbürden, die keine dauernde Reaktionen im sub-
Sekundenbereich erfordern. Selbst Steuerreaktionen im Sekunden-
raster würden jeden Menschen überfordern und sollten durch das
Rechnersystem wahrgenommen werden. Verläßlichkeit heißt, daß sich
ein Anwender bei der Lösung seiner Probleme auf die vom System angebotenen Dienste verlassen können muß, also der zu erbringende
Dienst möglichst unanfällig gegen Systemfehler sein soll.
Gerade in der Prozeßvisualisierung muß sichergestellt werden, daß
wichtige Ausgaben auch tatsächlich bis zum steuernden Menschen gelangen. Dies heißt, daß die Sichtbarkeit für bestimmte Bildschirm-
Ausgaben garantiert werden muß. Aus der Anforderung, nach Echtzeitfähigkeit folgt, daß die Bearbeitungsreihenfolge von Aufträgen
durch den X Server vom Anwender bestimmt werden können muß. Nur so
können die Reaktionszeiten für einen bestimmten Auftrag garantiert
werden. Das X Window System erlaubt überlappende Fenster. Es muß
aber sichergestellt werden, daß ein Fenster, in das eine wichtige
Ausgabe erfolgt, nicht gerade durch ein anderes Fenster verdeckt
wird.

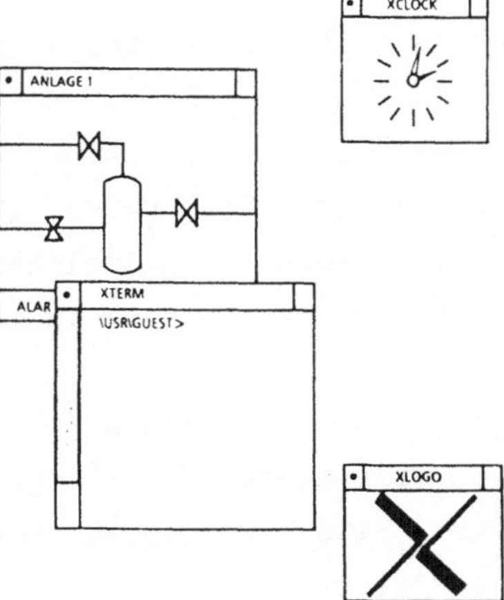

Bild 2: Überlappung von Alarmfenstern

Prinzipiell läßt sich das gewünschte Verhalten auch erreichen, indem ein X Server ausschließlich mit einer Anwendung kommuniziert,
hierdurch bestimmt die Anwendung selbstständig, welche Aufträge in
welcher Reihenfolge abgegeben werden. Weiterhin wird ein solcher

Ansatz häufig auch gänzlich ohne den Einsatz eines Window-Managers realisiert. Hierdurch kann (und **muß**) die Anwendung selbst dafür sorgen, daß wichtige Informationen auch tatsächlich am Bildschirm sichtbar sind. Eine solche vollständige Kontrolle über den Bild-schirm durch ein Applikationsprogramm wird jedoch nur dann sinn-voll sein, wenn lediglich eine Applikation auf einem Rechner Graphikausgaben auf einen Bildschirm macht, also für geschlossene Systeme. Hierdurch geht jedoch einer der Vorteile des X Window Systems verloren nämlich die Möglichkeit, daß mehrere Anwendungen, die auch auf mehreren Rechnern verteilt ablaufen können, gleich-zeitig Ausgaben auf einem Bildschirm machen. Zudem kann man mit einem solchen Vorgehen keinen Mischbetrieb zwischen Standard X Window Applikationen und Anwendungen der Prozeßvisualisierung er-reichen. Eine weitere Eigenschaft, die für Prozeßvisualisierungs-Anwendungen erforderlich ist, ist daß das Blinken von Bildteilen auf eine möglichst komfortable Art unterstützt werden sollte. Aus ergonomischen Gesichtspunkten sind mindestens drei unter-schiedliche Blinkfrequenzen und mehrere Blinkfarben notwendig. Wie kann man nun also eine offene, verteilte Prozeßvisualisierung mit Hilfe des X Window Systems ermöglichen? Ein Vorbild hierfür bietet der Ansatz, der bei der Implementierung von SORIX zugrunde-liegt. Bei SORIX, dem Echtzeit UNIX Betriebssystem von Siemens können Standard UNIX Anwendungen und Echtzeit-Anwendungen ko-existieren. Hierbei erhalten die Echtzeitanwendungen den Vorrang gegenüber den Standard-UNIX-Anwendungen, um ein deterministisches Verhalten zu erreichen. Die Standard-UNIX-Applikationen werden durch dieses Vorgehen nicht gestört, da sie eine Deterministik ohnehin nicht vorraussetzen. Erreicht wird bei SORIX eine Auf-teilung in den Echtzeit- und Standard-UNIX-Bereich unter anderem über die Unterstützung von Prioritäten. Was liegt näher, als diesen Ansatz auch für das X Window System für SORIX zu verfolgen? Führt man auch für Windows Prioritäten ein, so kann man das ge-forderte deterministische Verhalten über diese Prioritäten erreichen. Windows niedriger Priorität dürfen Windows höherer Priorität nicht überlappen. Ausgaben in Windows werden nach der Windowpriorität bearbeitet. Durch die Vergabe von Prioritäten an Windows lassen sich auch verteilte Anwendungen auf einem Bild-schirm koordinieren. Eine Verarbeitung von Prioritäten erfordert sowohl Änderungen im X Server und Erweiterungen der X Library, als

auch die Entwicklung eines speziellen Window Managers. Es muß nämlich der X Server dergestalt erweitert werden, daß er Window-Re-
sourcen nach Priorität verwaltet und Aufträge für Windows nach der
jeweiligen Priorität durchführt. Zusätzliche X Library Aufrufe
müssen geschaffen werden, die es einer Anwendung erlauben, die
Priorität eines Windows festzulegen und abzufragen. Schließlich
muß ein spezieller Windowmanager bei Vergrößerung und Verschiebung
von Fenstern die Priorität berücksichtigen. Diesem Vorgehen kommt
eine Eigenschaft des X Window Systems zugute, nämlich die von
vorne herein berücksichtigte Erweiterbarkeit. Über einen **Standard-
Aufruf** kann man die von einem spezifischen X Server unterstützten
Erweiterungen ermitteln. Hierdurch läßt sich sicherstellen, daß
man sich an der Schnittstelle nicht allzu weit vom Standard X
Windows entfernt. Legt eine Anwendung nun nicht explizit die
Priorität eines Windows fest, so erhält das Window die niedrigste
Priorität. Hiermit läßt sich eine Koexistenz von "Echtzeit
Windows" für Anwendungen der Prozeßvisualisierung und normalen
Windows für Standardclients erreichen. Erweitert man nun den
Windowmanager noch um die Möglichkeit, interaktiv die Priorität
von Windows auf dem Bildschirm zu setzen und zu verändern, so las-
sen sich verteilte Mischanwendungen komfortabel administrieren und
auf die spezifischen Anforderungen der Anwendung zurechtschneiden.
Die zweite Anforderung für die Prozeßvisualisierung, das Blinken ,
wird auch über ein Zusammenspiel von XServer-Erweiterung, XLibrary
Erweiterung und Hardware Hilfsmittel realisiert. Ein Blinken
könnte man auch allein mit Hilfe der XLibrary realisieren. Ein
Programm könnte z.B. zyklisch einen Farbwert der "Color Lookup
Table", also der Instanz, mit deren Hilfe Pixelwerten Rot-, Grün-
und Blau-Anteile, also die eigentliche Farbe zugeordnet wird, von
einer Farbe in eine andere ändern. Eine solche Vorgehensweise
hätte jedoch gravierende Nachteile. Einerseits würde die Blinkfrequenz in Abhängigkeit von der Belastung des XServer und des
eventuell zwischen ihm und der Anwendung liegenden LAN variieren,
da ja, wenn der XServer viele andere Aufträge ausführen muß, der
Auftrag zum Umhängen der lookup table erst später drankommt, als
wenn die Belastung des Servers gering ist. Zudem puffert die
XLibrary intern Aufträge und übermittelt sie in Paketen, was dazu
führt, daß eine Applikation bei Abgabe eines "Blinkauftrags" den
Zeitpunkt der Übermittlung dieses Auftrages nicht genau bestimmen

kann. Ein Mensch erkennt eine ungleichmäßige Blinkfrequenz jedoch
genau und empfindet dies als unangenehm. Weiterhin werden durch
das zyklische Abgeben von Aufträgen sowohl der Applikationsrechner
als auch das Netzwerk unnötig belastet. Der SIMICRO SX verändert
ähnlich dem Beispiel zyklisch Werte der Colour Lookup Table, jedoch nicht auf der Applikationsseite, sondern durch den XServer
selbst. Über einen zusätzlichen Aufruf der Xlibrary kann die
Applikation für bestimmte Indizes der Colour Lookup Table Blinkfrequenzen und Blinkfarben definieren. Die Auswahl dieser Indizes
kann durch die Applikation beliebig bestimmt werden. Nachdem eine
oder mehrere "Blinkfarben" definiert wurden, kann diese Farbe über
die normalen Mittel der XLibrary bestimmten Objekten zugeordnet
werden, die danach völlig ohne Zutun der Applikation in der gewünschten Frequenz und Farbe blinken.
Mit den beschriebenen Aspekten meine ich lassen sich Prozeß-
visualisierungen realisieren, die keine Abstriche bei den Echtzeitanforderungen hinnehmen müssen, ohne daß man vom Standard
XWindows zuweit abweichen muß. Ein Mischbetrieb von Echtzeit-
anwendungen mit Standardclients ist möglich, da die jeweiligen Erweiterungen völlig transparent für die Anwendungen realisiert
werden. Der SIMICRO SX mit SORIX bietet also auch bei der Realisierung seiner Graphik eine gelungene Synthese von Standard-UNIX-
Mitteln und den nötigen Erweiterungen für die Echtzeitanwendungen.

Hard- und Software für X Windows beim SIMICRO SX

Für den SIMICRO SX werden auf der Client-Seite sowohl die X
Library, als auch darauf aufsetzende Toolkits wie das vom MIT
definierte X Toolkit und die Bedienoberfläche OSF/Motif angeboten.
Zusätzlich gehört eine ganze Anzahl von Standard-Applikationen zum
Lieferumfang der jeweiligen Pakete.
Auf der Server-Seite sind sowohl sogenannte X-Terminals, Work-
stations oder PCs, die das X Window System unterstützen, als auch
ein Grafik-Modul für den direkten Anschluß eines Graphik-Terminals
über X ansprechbar. Für besonders hohe Anforderungen an Auf-
lösung, Performance und Anzahl der Farben entsteht derzeit eine
Grafik-Erweiterungsbaugruppe im MBII-Format für den SIMICRO SX.
Bei X Terminals handelt es sich um speziell auf das X Window
System angepasste Terminals, die über LAN angeschlossen werden.

Als Firmware befindet sich auf dem Terminal sowohl TCP/IP, als
auch der X Server, sodaß hierdurch ein im Verhältnis zu Work-
stations preislich günstiges Ausgabemedium zur Verfügung steht,
das speziell auf die Anforderungen des X Window Systems ausgerichtet ist. Auf dem Markt sind sowohl Schwarz-Weiß-, als auch
Farb-X Terminals vorhanden. Ein gravierender Nachteil dieser
Standard-Terminals ist die mangelnde Eignung der derzeit am Markt
erhältlichen Produkte für die Automatisierung. Zum einen besitzen
sie als Standard-Komponenten keine der vorher beschriebenen Echt-
zeit-Erweiterungen. Zudem ist derzeit eine hohe Belastung der
Terminals problematisch, da häufig das Eröffnen eines zusätzlichen
Fensters bei Speicherknappheit auf dem Terminal zum Absturz des X
Servers führt, der ein Rücksetzen des X Terminals nötig macht.

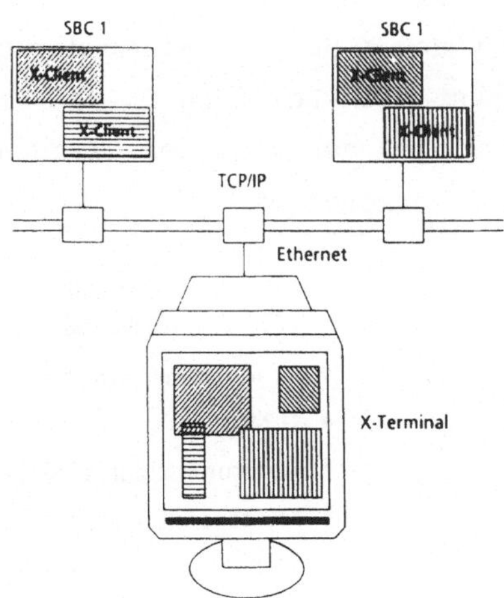

Bild 3: Grafik-Ausgabe über X-Terminals.
Für den Anschluß von Graphik-Bildschirmen an den SIMICRO SX steht
ein Grafikmodul zur Verfügung, welches speziell auf die Belange
des SIMICRO SX zugeschnitten ist. Über den auf dem Modul vorhandenen RGB-Stecker kann ein Bitmap-Farbterminal angeschlossen
werden, auf dem die Graphik-Ausgaben erfolgen. Die Tastatur und
optomechanische Maus wird an eine serielle Schnittstelle des
Single Board Computers SBC, der beim SIMICRO SX als CPU-Baugruppe
dient, angeschlossen. Das Grafikmodul unterstützt Auflösungen von
bis zu 1024 X 768 Bildpunkten und kann gleichzeitig 16 Farben am

Bildschirm darstellen. Als Graphik-Prozessor wird der bekannte TI 34010 Graphik-Prozessor von Texas Instruments verwendet, der aufgrund seiner hervorragenden Eigenschaften besonders für Graphik-Anwendungen in Verbindung mit X Windows geeignet ist. Der X Server läuft auf dem Modul selbst unter Kontrolle des TI 34010 ab und entlastet hierdurch den SBC von den hierfür erforderlichen Aktivitäten. Auf dem Modul stehen 2 bzw. 4 MB lokales DRAM zur Verfügung. Im Hochlauf wird der X-Server auf das Modul geladen und kommuniziert mit seinem Host-Rechner, dem SBC, über eine spezielle Treiber-Schnittstelle. Sowohl in diesem Treiber, als auch in der Firmware des Grafikmoduls sind Funktionen zur Nutzung des Grafikmoduls als Systemconsole integriert, sodaß ein Betrieb ausschließlich mit dem Graphik-Terminal erfolgen kann. Zusammen mit dem Grafikmodul und der zugehörigen Software erfüllt der SIMICRO SX als "embedded System" die gleichen Aufgaben, wie ein X-Terminal bzw. eine Workstation mit dem X Server. Einer der großen Vorteile hiervon ist, daß die Workstation-Leistung in das Prozeßsteuerungssystem eingebettet ist. So gesehen wird der SX zusammen mit dem Grafikmodul zur "embedded Workstation".

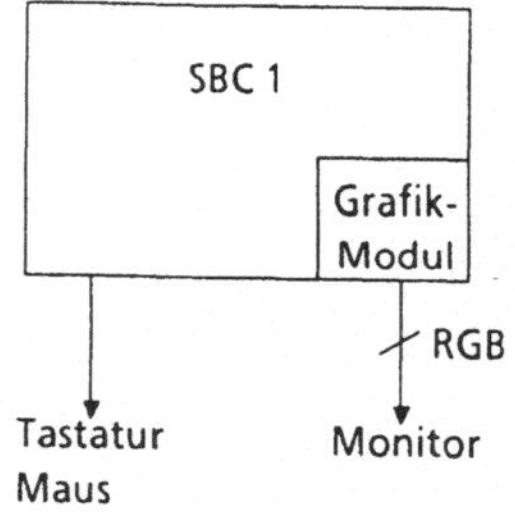

Bild 4 : Graphik-Ausgabe über das Graphikmodul XGM1.
Für höhere Ansprüche an Auflösung und Farben wird die Graphik-Erweiterungsbaugruppe XBS3 entwickelt. Diese ist eine Erweiterungsbaugruppe des SIMICRO SX Systemes, auf der die Graphik integriert wurde, bietet also einen Steckplatz für Kommunikationsmodule und 2 MB EPROM. Zusätzlich läuft, ähnlich dem Grafikmodul XGM1, der X Server auf der Baugruppe ab. Die Kommunikation zum Betriebssystem übernimmt wie beim XGM1 ein Statthalterprozeß. Der X Server und die Graphik-Funktionen werden vom Graphikprozessor TI 34020 ausgeführt, der optional durch den zugehörigen Gleitkommaprozessor TI

34082 ergänzt wird. Die Baugruppe erlaubt Auflösungen von
1280x1024 und 640x480 Bildpunkten, bei denen Bildwiederhol-
frequenzen von 75 und 80 Hz erreicht werden (non interlaced). Der
Bildspeicher ist 2 MB groß, der Arbeitsspeicher auf der Baugruppe
kann 2 bis 16 MB groß sein. Es sind gleichzeitig 256 aus 4096
Farben darstellbar. Auch die XBS3 unterstützt als Server-Erweite-
rung Blinken. Zusätzlich zum RGB-Anschluß für den Bildschirm kann
man auf der XBS3 eine PS2-kompatible Maus und Tastatur lokal anschließen. Der Zugriff auf den Speicher der XBS3 erfolgt aufgrund
der besonderen Möglichkeiten des TI 34020 "Memory mapped", das
heißt, der gesamte Speicher der Erweiterungsbaugruppe läßt sich
von SORIX aus über die normalen Speicher-Speicher Transfers an-
sprechen. Hierdurch erhält man eine besonders performante
Kommunikationsschnittstelle, die auch für den Transfer von großen
Datenmengen geeignet ist. Auch die XBS3 erlaubt über eine Konsol-
terminal-Emulation den Betrieb des SIMICRO SX ohne zusätzliche
Terminals nur über das Graphik-Terminal.

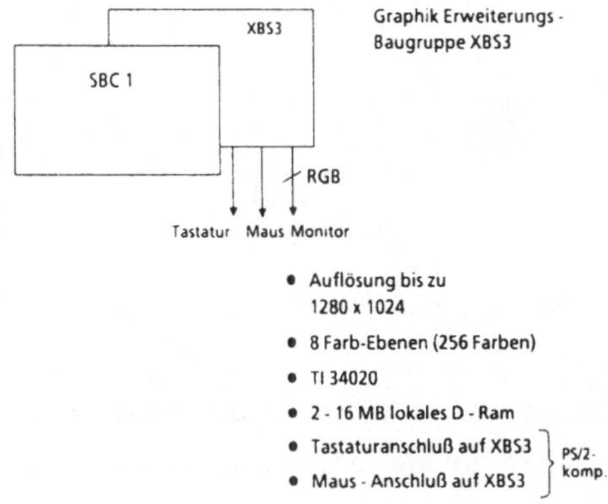

Bild 5 : Graphik-Ausgabe über Erweiterungsbaugruppe XBS3
Sowohl für XGM1 als auch für XBS3 wird eine Software-Umschaltung
von Tastatur und Maus auf verschiedene SBCs eines MBII-Systems ermöglicht. Hierdurch ist speziell im Warten-Betrieb die Bedienung
mehrerer Bildschirme über eine Maus und Tastatur möglich. Maus-
und Tastaturtelegramme werden hierbei über den Rückwandbus MBII
vom System, an dem sie physikalisch angeschlossen sind an das
System geschickt, das derzeit bedient werden soll.
Mit Hilfe der beschriebenen Hardware und Software lassen sich auf
besonders flexible Weise verteilte Visualisierungsanwendungen
realisieren. Dies wird besonders für Anwendungen im Zuge des Zu-

sammenarbeitens mehrerer SBCs eines Baugruppenträgers wichtig, also für Multicomputing-Anwendungen. Als Erweiterung der lokalen und der TCP/IP Sockets gibt es beim SIMICRO SX nämlich auch am MBII Sockets. Diese lassen sich für Graphik-Applikationen basierend auf dem SX X Window System derart nutzen, daß an einen SBC als Graphik-SBC des Graphik-Modul physikalisch angeschlossen ist, jedoch auf den anderen SBCs am MBII weitere Applikationen Ausgaben auf diesen Graphik-Bildschirm durchführen können.

Mit Hilfe der beschriebenen Hard- und Software-Mittel lassen sich wie in **Bild 6** verdeutlicht Konfigurationen sowohl für lokales als auch für am LAN oder MBII verteiltes Bedienen und Beobachten realisieren.

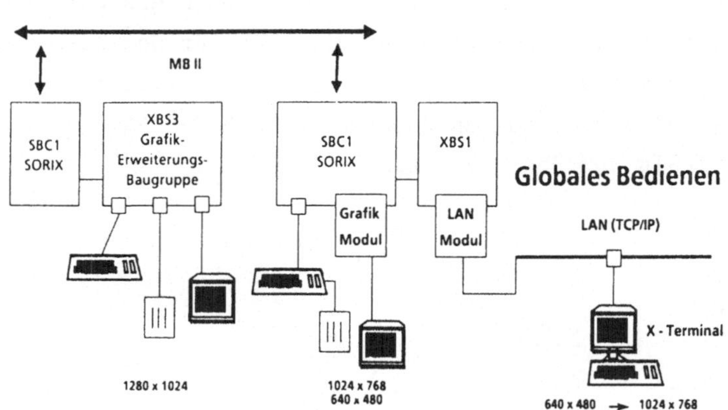

Bild 6: Konzepte für Bedienen und Beobachten für SIMICRO SX

XMove, das objektorientierte Visualisierungssystem des SIMICRO SX
Speziell für den SIMICRO SX wurde bei Siemens ein System zur komfortablen Erstellung von Anwendungen zur Prozeßvisualisierung entwickelt mit dem Namen XMove[4]. XMove dient dazu, den Anwender von der relativ aufwendigen Programmierung von Prozeßvisualisierungsanwendungen direkt auf der X Lib zu entlasten. Zusätzlich zur Möglichkeit, Bilder interaktiv zu editieren, bietet XMove speziell für die Prozeßautomatisierung geeignete Hilfsmittel.

---

[4] XMove bedeutet Meter Object Visualizing and Editing Tools

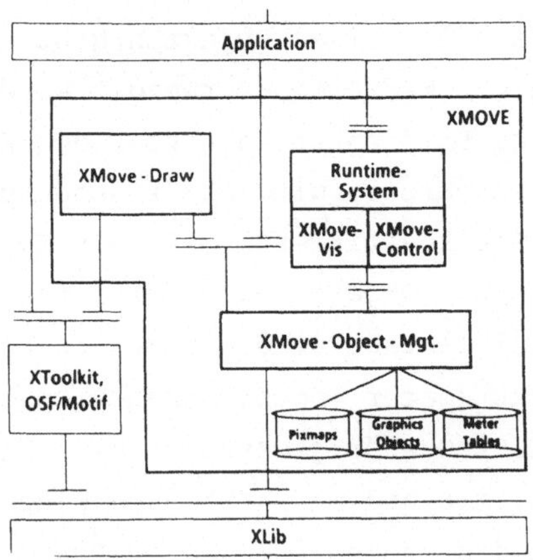

Bild 7: XMove Systemaufbau

Das Visualisierungspaket XMove besteht unter anderem aus dem
vektororientierten Editor XMove-Draw, mit dem man, ähnlich wie bei
anderen Graphikeditoren, aus graphischen Grundelementen wie Kreis,
Elipse, Polyline etc. Bilder konstruieren kann. Zusätzlich besteht
die Möglichkeit, die in X Window häufig verwendeten Pixmaps über
einen speziellen Pixmap-Editor zu erstellen und diese als Elemente
in vektororientierten Bildern zu verwenden. Teile eines so erstellten Bildes können zu Objekten zusammengefasst und mit einem
Namen versehen werden. Programmen steht eine Library zur Verfügung
über die mit den so definierten Bildern und Bildobjekten
Operationen durchgeführt werden können. Über ein Laufzeitsystem
kann man unter anderem Bilder anwählen und Bildobjekte verschieben, vergrößern und Farbänderungen für bestimmte Objekte
durchführen.

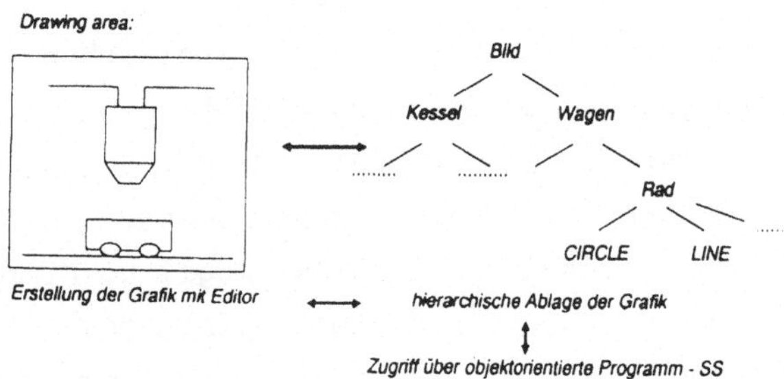

Bild 8: Objektorientierte Graphik

Zusätzlich zu diesen allgemeinen Objekten gibt es Meterobjekte, die zur Darstellung von Meßwerten verwendet werden können. Ein mit dem Editor erstelltes Bildobjekt kann über den Editor zum Meterobjekt gemacht werden. Hierzu wird das zugehörige Meßinstrument über den Graphikeditor konstruiert, z.B. als Meßuhr mit rotierendem Zeiger und als Meter-Objekt deklariert. Für eine anzuzeigende Meßgröße z.B. einen Füllstand wird hierzu ein Name gewählt, unter dem das Meßinstrument später angesprochen wird z.B. FÜLL1. Diesen Namen erhält auch das zugehörige Meterobjekt. Mit dem Objekt wird eine Meßwerttabelle, min und max-Werte und eine Farbliste für Farbumschläge durch die zugehörige Editor-Funktion definiert und verbunden. Zur Laufzeit übergibt das Programm lediglich die darzustellenden Werte per Aufruf an das Meterobjekt, welches selbständig die graphische Darstellung, Skalierung und eventuell notwendige Farbwechsel vornimmt. Hierdurch wird das Anwenderprogramm weitgehend von den für die graphische Darstellung der Meßwerte notwendigen Operationen entlastet.

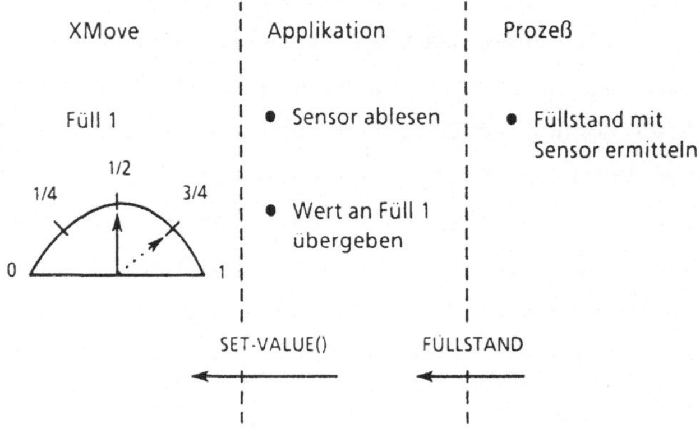

Bild 9: Beispiel für Meterobjektschnittstelle
Da ja die übergabe der Meßwerte über einen symbolischen Namen erfolgt, werden das Programm, das die Meßwerte übergibt und die eigentliche Meßwertdarstellung entkoppelt. Eine Änderung des graphischen Erscheinungsbildes eines Meßinstrumentes ist ohne Änderung beim den Meßwert liefernden Programm möglich. Hierdurch werden auch Möglichkeiten für "Rapid Prototyping" geschaffen, da eine Prozeßdarstellung in seiner optischen Erscheinung schon über ein Dummy-Programm, welches die Meßwerte lediglich aus einer Datei liest, betrachtet werden kann. Als Typen von Meterobjekten sind als einwertige "digital" (Ausgabe des Zahlenwertes), "symbol"

(Ausgabe einer Folge von Objekten je nach Größe des Meßwertes) und "hand" (Rotierende Zeiger, Balkendiagramme,...) und als mehr-wertige "table" (Ausgabe einer Wertetabelle) und "curve" (Ausgabe einer Kurve über Spline, Polyline oder Balken) realisiert. Ein-dimensionale Objekte für Werteeingaben wie Schieberegler oder Fel-der können ebenfalls als spezielle Meterobjekte definiert werden. Da XMove speziell für die Verwendung des X Window Systemes ent-wickelt wurde, ist auch ein Mischen von Standard Toolkits wie OSF/Motif und XMove möglich. Hierdurch kann der Anwender z.B. eine Programmbedienung über Standard Motif Widgets und die Meßwert-anzeige über XMove im gleichen Bild realisieren.

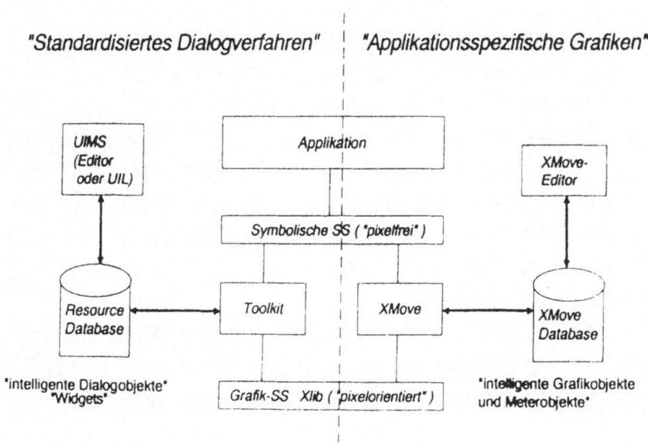

Bild 10: Mischbetrieb OSF/Motif XMove

VMEexec - Implementierung eines Echtzeitkern-Schnittstellen Standards

JuttaGöhringer

Motorola GmbH

Geschäftsbereich Computersysteme

Postfach 820960, 8000 München 82

Tel.: 089-92103-293 Fax: 089-92103-101

Für den Echtzeitbereich gibt es noch keinen Softwarestandard. Dies ist aber die Voraussetzung für kompatible Softwareprodukte und die Möglichkeit für die Anwender aus dem großen Softwareangebot die optimalen Produkte für ihre Applikation zusammenzustellen. Die Firma Motorola hat den ersten Schritt getan, indem sie vor 2 Jahren die "Real-Time-Executive-Interface-Definition (RTEID)"-Spezifikation entwickelt hatte. Diese Spezifikation wurde dem VITA Software Subcommitee als Vorschlag für einen Echtzeitkern-Schnittstellenstandard eingereicht. Die Version die dann von dem VITA Gremium ausgearbeitet wurde heißt ORKID (Open-Realtime-Kernel-Interface-Definition). Die erste Version wurde im November 1989 vorgestellt, die 2. Version gibt es seit Januar 1990.

Da es einige Zeit dauert bis ein Standardisierungsprozeß abgeschlossen ist, hat Motorola ihr neues Echtzeit-Betriebssystemprodukt "VMEexec" auf RTEID aufgebaut. Sobald die endgültige Version von ORKID vorliegt, wird VMEexec an ORKID angepaßt werden.

VMEexec ist ein Softwarepaket, das die Entwicklung von Echtzeitapplikationen unter Unix ermöglicht. Die Echtzeitsoftware besteht aus einem RTEID-kompatiblen Echtzeitkern, SVIDlib, NETlib und diversen Gerätetreibern. SYSTEM V/68, die Unix System V Version von Motorola dient als Entwicklungsumgebung, kann aber auch für zeitunkritischen Aufgaben in der VMEexec Anwendung eingesetzt werden. XRAY68K, ein Source Level Debugger, ermöglicht Task-Debugging und System-Monitoring. Für die VMEexec Programmentwicklung gibt es Compiler für die Sprachen Assembler, C, Pascal und Modula 2 (Ada folgt).

Für den Echtzeitmarkt gibt es ein großes Produktangebot von vielen verschiedenen Herstellern. Die Anwender haben die nicht immer leichte Aufgabe die richtigen Hardware- und Softwarekomponenten für Ihre Applikation auszuwählen. Aufgrund des VMEbuses ist es relativ einfach Produkte verschiedener Hardwareanbieter zu kombinieren. Im Softwarebereich hat sich UNIX durch seine weite Verbreitung praktisch als de-facto Betriebssystemstandard durchgesetzt. Für den Echtzeitbereich gibt es jedoch noch keinen Softwarestandard. Motorola hatte sich zum Ziel gesetzt diese Lücke zu schließen. Deshalb wurde eine Schnittstellendefinition für Echtzeitkerne entwickelt. Die erste Version der "RTEID (Real-Time-Executive-Interface-Definition)-Spezifikation wurde im September 1988 fertiggestellt.

Eine Schnittstellendefinition wird aber erst zum Standard, wenn sie von vielen Anwendern und Herstellern akzeptiert wird. Das setzt voraus, daß der Standard von einem unabhängigen Gremium entwickelt wurde. Deshalb hatte Motorola der VITA (VMEbus International Trade Association) RTEID als Vorschlag eingereicht. Innerhalb der VITA gibt es einen Unterausschuß Software (Software Subcommitee SWSC), der sich mit Schnittstellendefinitionen befaßt. Das aktuelle Projekt dieses Ausschusses ist ORKID, die VITA Version von RTEID. ORKID steht für Open-Real-time-Kernel-Interface-Definition. Diese Spezifikation wurde erstmals im November 1989 der Öffentlichkeit vorgestellt. Die aktuelle Version gibt es seit Januar 90 und kann bei der VITA angefordert werden.

Standardisierte Softwareschnittstellen bieten die Grundlage für kompatible Softwareprodukte und somit ein großes Software Angebot mit höherer Qualität. Die Anwender werden dadurch unabhängiger von den Softwareanbietern. Applikationssoftware läßt sich einfach in mehreren Projekten wiederverwenden, wodurch der Lernaufwand für neue Projekte reduziert wird.

Die RTEID-Spezifikation definiert die Schnittstelle zu einem modernen Echtzeitkern. Bei der Definition wurden die Erfahrungswerte der bis dahin existierenden Kerne berücksichtigt. Die Spezifikation läßt es dem Entwickler offen einen portierbaren, robusten und deterministischen Kernel-Code zu schreiben. Die RTEID ist Hardware unabhängig, das bedeutet sie schreibt keinen Prozessortyp vor. Sie definiert Objekte und Operationen, beziehungsweise Systemfunktionen, die vom RTEID-kompatiblen Echtzeitkern ausgeführt werden. Eine Zusammenfassung aller RTEID-Systemfunktionen ist in Tabelle 1 gegeben. Die Syntax einer Funktion gibt den Namen und die dazugehörigen Parameter an. Zum Beispiel ist die Funktion "Senden eines Events an eine Task" so definiert:

$$ev\_send\ (tid,\ event)$$

"tid" identifiziert die Task welche den oder mehrere Events erhalten soll. Mit "event" werden die gewünschten Events bitweise selektiert.

Ein RTEID-kompatibler Kern ist objektorientiert. Folgende Objekte werden durch die Spezifikation festgelegt:

Tasks, Memory Regions, Memory Partitions, Message Queues und Semaphoren.

Objekte können als lokale oder globale Objekte erzeugt werden. Lokale Objekte sind nur dem erzeugenden Kern bekannt. Bei einem globalen Objekt wird die Information über dieses Objekt, von dem erzeugenden Kern, an alle anderen Kerne in einem Multiprozessorsystem weitergegeben. Es ist Aufgabe des RTEID-kompatiblen Kernes die Verwaltung der lokalen und globalen Objekte durchzuführen. Zur Applikationsebene hin ist es völlig transparent wo sich ein Objekt befindet. Dieses Konzept unterstützt die Multiprozessoreigenschaft des RTEID-Kernes, wobei auch diese Definition völlig Hardwareunabhängig ist. Somit ist es dem Entwickler des Echtzeitkernes überlassen wie und über welches Medium die Multiprozessorkommunikation durchgeführt wird.

Die Funktionalität eines RTEID-kompatiblen Kernes entspricht weitgehend der eines ORKID-kompatiblen Kernes. RTEID bildet die Grundlage von VMEexec, das Echtzeit-Multiprozessor-Betriebssystem-Produkt von Motorola. Wenn ORKID in einer endgültigen Version vorliegt, wird VMEexec ORKID unterstützen.

VMEexec ist ein Softwarepaket, das die Entwicklung von Echtzeitapplikationen unter UNIX ermöglicht. Es ist die Implementierung des Software-Standardisierungskonzeptes von Motorola. Neben dem RTEID-kompatiblen Echtzeitkern beinhaltet VMEexec noch SVIDlib, NETlib und diverse Gerätetreiber. Bild 1 zeigt die Struktur der VMEexec Applikation.

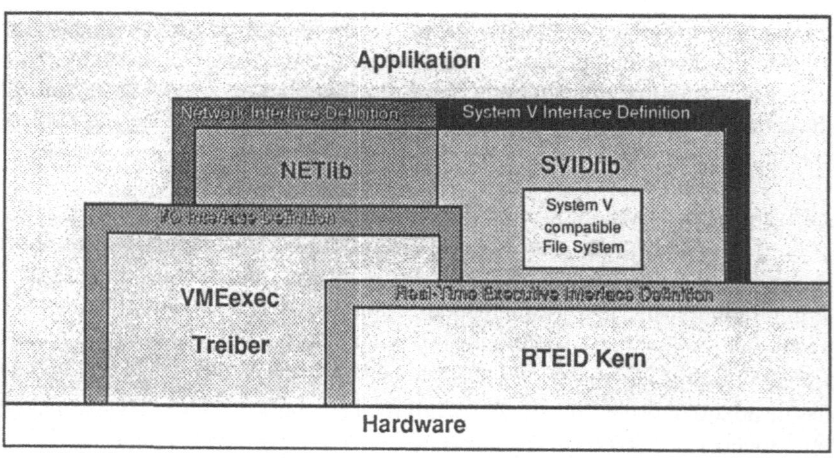

Bild 1: Struktur der VMEexec Applikation

Daraus kann man erkennen, daß die Applikation in verschiedene Softwarekomponenten aufgeteilt ist und diese durch definierte Schnittstellen verbunden sind. Die RTEID-Schnittstelle bildet die Basis der Applikation. Die Funktionsaufrufe an die VMEexec Gerätetreiber sind definiert und für alle Treiber gleich.

VMEexec Treiber haben eine vergleichbare Struktur mit SYSTEM V/68 Treibern. Sie sind in 6 Unterroutinen aufgeteilt: init, open, close, read, write, ioctrl und können eine Interrupt Service Routine enthalten. Die Treiber ermöglichen synchrone und asynchrone Bearbeitung von Ein/Ausgabeanforderungen.

SVIDlib, die "SYSTEM V Interface Definition"-Bibliothek bietet die UNIX-ähnliche Programmierschnittstelle in der Echtzeitapplikation. Alle Funktionen, die in einer Echtzeitumgebung Verwendung finden und nicht optimaler durch direkte Kernelaufrufe ausgeführt werden, wurden aus SVID in der SVIDlib implementiert. SVIDlib erspart einem UNIX-Programmierer nicht nur den Lernaufwand, sondern ermöglicht auch, daß Prozesse die SVIDlib-Aufrufe verwenden, ohne Änderungen sowohl in VMEexec als auch in SYSTEM V/68 ablaufen können.

Ein weiterer Vorteil von SVIDlib ist die Unterstützung von 2 Filesystemen. Das Standard SYSTEM V/68 File System und ein für Echtzeitanforderungen zugeschnittenes Fast-File-System. Dieses Filesystem hat dieselbe Struktur wie ein SYSTEM V/68 File System. Es unterscheidet sich dadurch, daß Files aus zusammenhängenden Diskblöcken bestehen und der Filezugriff direkt und ungepuffert erfolgt. Eine Task kann durch das Setzen einer SVIDlib-Variablen bestimmen, ob sie auf das Fast-File-System des Zielsystems oder auf das SYSTEM V/68 Filesystem im Host zugreifen möchte. Auch ein Prozess im Host kann auf das Fast-File-System im Zielsystem zugreifen, indem er die Anforderung von einer Servertask im Zielsystem ausführen läßt. Die Servertask kann leicht vom Anwender geschrieben werden.

Als letzte Schnittstelle in der VMEexec Applikation bleibt noch die Oberfläche zur **NETlib** zu erwähnen. Sie besteht aus der AT&T BSD 4.3 Socket Schnittstelle.

SYSTEM V/68, die UNIX System V Version von Motorola, dient als Software-Entwicklungsumgebung. Somit stehen dem Anwender die Standard UNIX-Tools für die Programmentwicklung zur Verfügung. Für die VMEexec Systemgenerierung und für die Host-Zielsystemkommunikation wurden zusätzliche Utilities geschrieben. SYSTEM V/68 kann aber auch in der VMEexec Anwendung für zeitunkritische Aufgaben eingesetzt werden. Prozesse, welche unter SYSTEM V/68 ablaufen, können mit Tasks im Echtzeitzielsystem über die Multiprozessorschnittstelle kommunizieren.

Die **Multiprozessorkommunikation** zwischen dem UNIX-Host und den Echtzeit-Zielsystemen oder zwischen den Zielsystemen kann über den VMEbus oder über das Netz (Ethernet) stattfinden. Folgende Funktionen können über Ethernet ausgeführt werden:
- Kommunikation zwischen verteilten Betriebssystemkernen,
- Booten, Laden und Debuggen,
- Alle "Tools" für die Multiprozessorkommunikation,
- SVIDlib Zugriff auf Host-Dienstleistungen.
Das verwendete Protokoll für die Ethernetkommunikation ist TCP/IP.

VMEexec wurde von Spezialisten entwickelt. Motorola hat sich auf die UNIX Komponenten, die Unterstützung der Motorola VME-Karten, die Netzwerkimplementierung und die Multiprozessorinterface-Software konzentriert. Der RTEID-kompatible Echtzeitkern pSOS+ wurde von der Firma Software Components Group geschrieben, die mit ihrem pSOS bereits eingehende Erfahrungen sammeln konnte. Der VMEexec Debugger ist der XRAY68K Debugger von der Fa. Microtec Research.

pSOS+ ist der erste RTEID kompatible Echtzeitkern und unterstützt alle CPUs der Motorola Prozessorfamilie MC680x0, sowie den MC88100 RISC Prozessor.
VMEexec enthält die Singleprozessorversion pSOS+ und die Multiprozessorversion

pSOS+m. pSOS+(m) unterstützt CPUs mit oder ohne MMU (Memory Management Unit), ist ROMfähig und ermöglicht Multiprozessing und Multitasking. Es wurde positionsunabhängig geschrieben und kann so überall im Speicher, sei es in RAM oder in ROM, plaziert werden. pSOS+(m) ist ein extrem schneller, robuster und hochentwickelter Multi-Tasking Echtzeitkern. Er unterstützt alle im RTEID spezifizierten Systemfunktionen (siehe Tabelle 1). Sie können sowohl in Assembler als auch in "C" programmiert werden.

pSOS+(m) wurde entwickelt um den höchsten Leistungsanforderungen gerecht zu werden. Jede zeitkritische Operation und jeder zeitkritische Pfad im Kern wurden optimiert, so daß nicht nur die maximale Ausführungszeit, sondern auch konstante Ausführungszeiten garantiert werden können. Dies zeigt sich in der "worst case interrupt response"-Zeit von 7 µs und der 19 µs konstanten "Context Switch"-Zeit. Gemessen wurden diese Werte mit einer MC68020 CPU bei 25 MHZ.

Der **XRAY68K** Debugger von der Fa. Microtec Research wurde auf die Motorola Delta Systeme portiert und für VMEexec angepaßt. Er wird auf dem Host-Prozessor ausgeführt und unterstützt Multiuser, Multitasking und Multiprozessor-Betrieb im Zielsystem, mit dem er über die Multiprozessorschnittstelle kommuniziert. Er kann "C" und Assembler Code bearbeiten und bietet eine benutzerfreundliche Bedienoberfläche durch seine "Multi-Window"-Eigenschaft. In den einzelnen Fenstern werden verschiedene Informationen (Code, Register, Variable, etc.) angezeigt. Zusätzliche Fenster können vom Anwender definiert werden. Er kann Variable lesen und ändern, oder im Assembler Mode Speicherzellen lesen und modifizieren, Einzelschritte auf Assemblerzeilen,"C"-Codezeilen oder "C"- Funktionen bearbeiten und den Systemstatus abfragen. Das waren nur einige Beispiele für die Möglichkeiten die XRAY68K bietet.

Für die Programmentwicklung stehen dem Anwender mehrere **Compiler** zur Verfügung:
Das VMEexec Software Paket enthält den Motorola Assembler und die SYSTEM V/68 "C"-Compiler für die Prozessoren MC68010, MC68020 und MC68030, sowie den Cross-Compiler für den MC88100.
Der Greenhill "C" und ABSOFT Fortran Compiler für die MC680X0 Familie wurden für VMEexec angepaßt. Sie sind über Motorola beziehbar.
Die Firma ACE(Associated Computer Experts bv) hat ihre EXPERT Compiler auf VMEexec portiert. Folgende Compiler werden von ACE für VMEexec angeboten: "C", Fortran 77, Modula-2 und Pascal. Sie erzeugen Code für die MC68010, MC68020, MC68030 Prozessoren, für die dazugehörigen Floating-Point-Coprozessoren MC68881 und MC68882 und unterstützen die MC68851 MMU für den MC68020 Prozessor sowie die MMU des MC68030 Prozessors.
An einer Portierung von ADA für VMEexec wird gearbeitet.

Entwicklungsumgebung

Die MVME147 CPU-Karte wurde als Host-CPU ausgewählt. Sie besitzt eine SCSI-Schnittstelle, was den Vorteil bringt, daß die Entwicklungssystemsoftware (SYSTEM V/68) vollständig über den SCSI-Bus arbeitet und somit der VMEbus für die Multiprozessorkommunikation frei bleibt.
Der Entwickler kann von einem SYSTEM V/68 Terminal aus die Echtzeitapplikation

entwickeln, compilieren und mit den Echtzeitkomponenten durch "vmexgen" verbinden. Dann kann das Objektmodul entweder über den VMEbus oder Ethernet ins Zielsystem geladen und dort gestartet werden. Tasks können auch nachträglich, also dynamisch ins Zielsystem nachgeladen und gestartet, laufende Tasks angehalten, wieder gestartet oder gelöscht werden. Vom gleichen Terminal aus wird der XRAY68K Debugger aufgerufen, der über die Multiprozessorschnittstelle ins Zielsystem eingreift.

Die Firma Philips in Eindhoven hat die Entwicklungsumgebung für VMEexec auf den PC portiert. Die Unterstützung für andere Hosts, z.B. SUN sind in Bearbeitung.

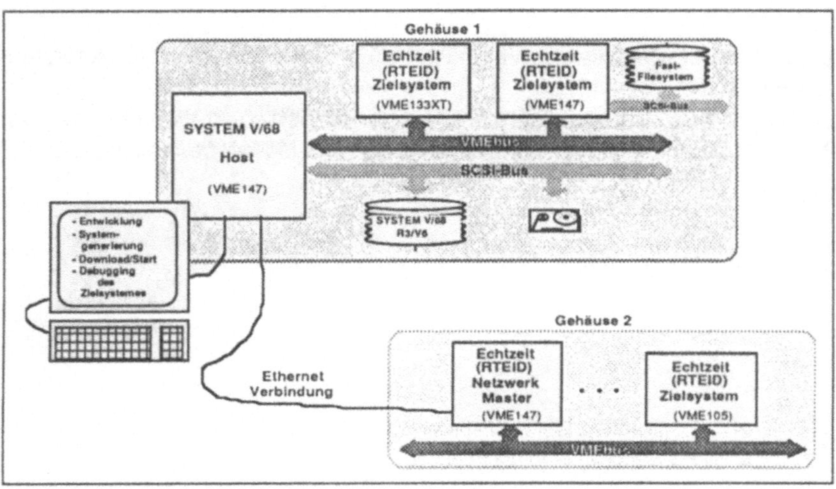

Bild 2: Beispiel für eine typische Betriebssystemumgebung von VMExec

Zielsystem-Hardware

VMEexec unterstützt die MC68010, MC68020, MC68030 Prozessoren auf folgenden Motorola Karten als Zielsystem-CPUs:
Die Familien MVME10X, MVME133 und MVME147, sowie die CPU-Karten MVME134, MVME135, MVME136, MVME141 und MVME143.
Der MC88100 Prozessor wird auf der MVME181- und MVME188-Karte unterstützt.
VMEexec für den MC68040 wird es mit einer späteren Version geben.

Das VMEexec Treiber Paket enthält Treiber für folgende Motorola VME-Karten:
• Serielle und parallele Schnittstellentreiber für die
  MVME10X, MVME133 und MVME147 Familien, die MVME134, MVME135,
  MVME136, MVME141, MVME143, sowie die MVME181 und MVME188 CPU-Karten.
• Serielle und parallele Schnittstellentreiber für die I/O Kontroller-Karten
  MVME335 und MVME332XT.

• Treiber für die Massenspeicherkontroller
  MVME320B, MVME321, MVME327, sowie für die SCSI-Schnittstelle auf der
  MVME147 CPU.
• RAM Disk-Treiber für die Anwender die ohne Massenspeicher auskommen wollen.
• Treiber für die Graphik-Kontroller-Karten MVME393 und MVME395.

Motorola hat zu den speziellen Treibern noch Treiberbibliotheken geschrieben. In
diesen Bibliotheken sind alle Funktionen zusammengefaßt, die für einen Gerätetyp
(serielle Schnittstelle, Disk, etc.) gleich und somit vom Gerät unabhängig sind. Ein
spezieller Treiber kann auf diese Bibliotheken zugreifen und muß nur noch den hardwareabhängigen Code enthalten. Damit läßt sich der Aufwand für die Treiberentwick-
lung verringern.

Mit jedem neuen Release von VMEexec werden Treiber für weitere Motorola Karten
hinzukommen.
Von der Fa. EBV gibt es zusätzliche Treiber für verschiedene I/O-Karten von XY-
COM, Motorola und National Instruments für die Analogdatenerfassung und IEC-Bus-
steuerung.

Tabelle 1: RTEID-Systemfunktionen

Task Direktiven

t_create		Erzeugen einer Task
t_ident		Anfordern der ID Nummer
t_delete		Löschen einer Task
t_start		Starten einer Task
t_restart		Zurücksetzen und starten der Task
t_suspend	(r)	Suspendieren einer Task
t_resume	(r/i)	Starten nach dem Suspendieren
t_setpri	(r)	Setzen der Task Priorität
t_mode		Ändern des Task Modus
t_getreg	(r/i)	Taskregister lesen
t_setreg	(r/i)	Taskregister setzen

Message Direktiven

q_create		Erzeugen einer Message Queue
q_ident		Anfordern der ID Nummer
q_delete		Löschen einer Queue
q_send	(r/i)	Senden einer Message an eine Task
q_urgent	(r/i)	Senden einer eiligen Message
q_broadcast	(r/i)	Senden einer Message an alle Tasks
q_receive	(r/i)	Empfangen einer Message

Signal Direktiven

as_catch		ASR aktivieren (Signal empfangen)
as_send	(r/i)	Senden eines Signals
as_return		Verlassen der ASR

Event Direktiven

ev_send	(r/i)	Senden eines Events
ev_receive		Empfangen eines Events

Semaphore Direktiven

sm_create		Erzeugen einer Semaphore
sm_ident		Anfordern der ID Nummer
sm_delete		Löschen einer Semaphore
sm_p	(r/i)	P-Operation
sm_v	(r/i)	V-Operation

Interrupt Service Direktive

i_return	(i)	Verlassen der ISR

Fatal Error Direktive

k_fatal		Aufruf der Fatal Error Routine

Time Direktiven

tm_set	(i)	Zeit und Datum setzen
tm_get	(i)	Anfordern von Datum und Zeit
tm_wkafter		Aufwecken nach einem Zeitintervall
tm_wkwhen		Aufwecken zu einer bestimmten Zeit
tm_evafter		Senden eines Events nach einem Zeitintervall
tm_evwhen		Senden eines Events zu einer bestimmten Zeit
tm_cancel		Ausschalten von Timer Events
tm_tick	(i)	Ankündigung eines Timer Ticks

Region Direktiven

rn_create		Erzeugen einer Region
rn_ident		Anfordern der ID Nummer
rn_delete		Löschen einer Region
rn_getseg		Anfordern eines Segmentes
rn_retseg		Zurückgeben eines Segmentes

Partition Direktiven

pt_create		Erzeugen einer Partition
pt_ident		Anfordern der ID Nummer
pt_delete		Löschen einer Partition
pt_getbuf	(r/i)	Anfordern eines Buffers
pt_retbuf	(r/i)	Zurückgeben eines Buffers
pt_sgetbuf	(i)	Anfordern eines physikalischen Buffers

Memory Direktiven

mm_l2p	Logische zu physikalische Adressumwandlung
mm_p2l	Physikalische zu logische Adressumwandlung
mm_pmap	Physikalisches "mappen"
mm_unmap	"unmappen"
mm_pread	Physikalisches Lesen
mm_pwrite	Physikalisches Schreiben
mm_sprotect	Supervisor-Adressbereich schützen

Dual-ported Memory Direktiven

m_ext2int	Adressumwandlung extern zu intern
m_int2ext	Adressumwandlung intern zu extern

("r" kennzeichnet "Remote pSOS+ Calls"
"i" markiert die Aufrufe die in einer ISR erlaubt sind).